T0029295

"The merciless seesaw of [Seager's] grief makes for harrowing reading. . . . Her story gleams with insights into what it means to lose a partner in midlife, and just as the widows helped Seager feel less alone, her story is sure to help any readers grappling with a similar loss."
—ANTHONY DOERR, *The New York Times Book Review*

"A singular scientist has written a singular account of her life and work."
—*Kirkus Reviews*, starred review

"It is the easiest thing in the world to resign yourself to what is, to curl around yourself and your circumstances. This is a book filled with hope and wonder, because falling in love after loss is a defiant act of optimism, much like searching the stars for life beyond our own little planet. You'll leave this book feeling possibility and inevitability, comforted by the knowledge that even in the dark, we are not alone."
—NORA MCINERNY PURMORT, author of *It's Okay to Laugh*

"Sara Seager's exploration of outer and inner space makes for a stunningly original memoir. Far from being dwarfed by the scale of exoplanets and galaxies, her most human tale of love, loss, and redemption is illuminated and given meaning by this backdrop. . . . A beautiful and compelling read."
—ABRAHAM VERGHESE, author of *Cutting for Stone*

"I absolutely loved this book. It presents both cutting-edge science and the deeply human side of a MacArthur award–winning woman astrophysicist. While searching for other planets in the universe, she grieves for her husband who died of cancer."

—TEMPLE GRANDIN, author of *Thinking in Pictures* and *The Autistic Brain*

"The miracle of this breathtaking book is the way Sara Seager's search for life in the universe mirrors her search for a fitting life here on Earth. Who knew that so much love and beauty and hope could come from so much confusion and fear and grief? Who knew that the macrocosm and the microcosm could end up being the very same thing?"

—MARGARET RENKL, author of *Late Migrations*

"Seager's beautifully written memoir strikes the perfect balance, weaving a richly told personal story with a clear and accessible tale of the birth and development of a new kind of astronomy—the search for other worlds like our own."

—KATIE MACK, author of *The End of Everything (Astrophysically Speaking)*

"This wondrous tale of discovery, loss, and love is both expansive and intimate."

—*Publishers Weekly,* starred review

"This thoughtful and affecting memoir of navigating life after loss reads like a comforting novel, inspiring others to follow their dreams and never give up on the possibilities of discovery and self-reflection."

—*Library Journal,* starred review

THE SMALLEST LIGHTS
IN THE UNIVERSE

THE SMALLEST LIGHTS
IN THE UNIVERSE

A MEMOIR

Sara Seager

CROWN
NEW YORK

LIBRARY OF CONGRESS CATALOGING-IN-PUBLICATION DATA
Names: Seager, Sara, author.
Title: The smallest lights in the universe / Sara Seager.
Description: First edition | New York: Crown, [2020]
Identifiers: LCCN 2020007803 (print) | LCCN 2020007804 (ebook) |
ISBN 9780525576266 (paperback) | ISBN 9780525576273 (ebook) |
ISBN 9780593238417 (international edition)
Subjects: LCSH: Seager, Sara. | Planetary scientists—Biography. |
Astrophysicists—Biography. | Widows—Massachusetts—
Concord—Biography. | Extrasolar planets.
Classification: LCC QB460.72 .S43 2020 (print) | LCC QB460.72 (ebook) |
DDC 523.4092 [B]—dc23
LC record available at https://lccn.loc.gov/2020007803
LC ebook record available at https://lccn.loc.gov/2020007804

Printed in the United States of America on acid-free paper

crownpublishing.com
randomhousebookclub.com

246897531

Title-page art: © iStockphoto.com

Book design by Dana Leigh Blanchette

FOR CHARLES

Author's Note

This is a work of nonfiction. To the best of my memory, everything that follows is true. Whenever possible, facts have been verified through secondary sources. A small number of names have been changed to protect the identities of private individuals.

THE SMALLEST LIGHTS
IN THE UNIVERSE

Not every planet has a star. Some aren't part of a solar system. They are alone. We call them rogue planets.

Because rogue planets aren't the subjects of stars, they aren't anchored in space. They don't orbit. Rogue planets wander, drifting in the current of an endless ocean. They have neither the light nor the heat that stars provide. We know of one rogue planet, PSO J318.5-22—right now, it's up there, it's out there—lurching across the galaxy like a rudderless ship, wrapped in perpetual darkness. Its surface is swept by constant storms. It likely rains on PSO J318.5-22, but it wouldn't rain water there. Its black skies would more likely unleash bands of molten iron.

It can be hard to picture, a planet where it rains liquid metal in the dark, but rogue planets aren't science fiction. We haven't imagined them or dreamed them. Astrophysicists like me have found them. They are real places on our celestial maps. There might be thousands of billions of more conventional exoplanets—planets that orbit stars other than the sun—in the Milky Way alone, circling our galaxy's hundreds of billions of stars. But amid that nearly infinite, perfect order, in the emptiness between countless pushes and pulls, there are also the lost ones: rogue planets. PSO J318.5-22 is as real as Earth.

There were days when I woke up and couldn't see much difference between there and here.

•

One morning it was only the distant laughter of my boys that persuaded me to push back the covers. Max was eight years old. Alex was six. They were looking out the window, their faces lit with kid joy. It was a blue-sky weekend in January, and a thin white blanket of snow had fallen overnight. Finally, a bright spot. We could go sledding, one of our family's favorite pastimes. After a quick breakfast, Max and Alex began putting on their snowsuits. With their plastic sleds stuffed into the car, we made the short drive to the top of Nashawtuc Hill.

The hill is a popular gathering spot in Concord, Massachusetts. It's steep and fast enough to thrill even grown-ups. It can get busy, but not that morning. There wasn't really enough snow to sled, and tall grass and weeds poked out of what snow was there. I tried to pretend for the sake of the boys that sledding would still be fun. I didn't believe it myself. I'd spent my entire life searching for lights in the dark; now I could see only the blackness that surrounded them. But we had gone to the trouble of getting to the top of the hill. The boys might as well try to get to the bottom.

There were two other women standing at the top, mothers talking and laughing with each other while their kids played. They were beautiful, their faces put together enough to make me resentful. I looked at them coldly. I thought: *Who gets up on a Sunday morning and thinks to do their makeup like that?* They looked like a picture from a brochure for happiness.

Max was big enough to get all the way down the hill. Even if he hit the weeds, he had enough mass and speed to pass over and through them. Physics weren't so much on Alex's side. He kept getting stuck. He tried going down a few times but eventually gave up. Seeing his brother hurtle to the bottom was too much for him to take. Alex sat there, pouting, right in the middle of the hill. He wasn't crying. He just spread himself across the hill and refused to move. If he wasn't going to have any fun, nobody was.

One of the women called over and asked if I could shift him. He was in the way, and she was afraid he was going to get hurt. I understood why he needed to be moved. I was also spent, my best plans undone. I wasn't in the mood to take orders from someone like her, from someone so pretty. I wasn't in the mood to take orders from anybody. I glared at her and shook my head.

She asked again.

"No," I said. "He has a problem."

She smiled and maybe even laughed a little. "Oh, okay," she said. "I mean, it's just that—"

I ignored her.

"It's just that the hill—"

"HE HAS A PROBLEM. MY HUSBAND DIED."

When you're in the ugly throes of grief, most people are repulsed by you. Nobody knows what to say or how to behave in your presence. Everybody's scared of what you represent, and in a way, I suppose, you learn to want them to be. The distance that people keep is a sign of respect: Your grief warrants a wide berth. You come to crave the ability to influence the movements of others, your sorrow a superpower, your sadness your most extraordinary trait. You come to crave the space.

I thought the woman on the hill would be shocked. I thought she would recoil. Instead, she did the strangest thing. She smiled, and then her eyes brightened. She became an oven, radiating warmth.

"Mine, too," she said.

I was stunned. I think I asked her how long she had been a widow. "Five years," she said. It had been only six months for me. *She's forgotten what it's like,* I thought. *How dare she laugh at me.*

I had an overwhelming urge to run, to return to my bed, lashed by my storms of molten iron, but Max was still having fun on the hill. It's moments like those, when you're torn in two, that you realize how alone you are. You need to find solutions to unsolvable problems. I decided that I'd take the boys home, and we'd get Alex the iPad. Then we'd come back. Alex could sit in the car and play, and Max could still sled. Hopefully the other widow would be gone by the time we got back.

She was still there when we returned. Meeting beautiful new people wasn't easy for me in the best of circumstances, and these were far from ideal. I had no idea what to do next. I tried to stand far away from her, to become even more repellent than I already felt. It didn't work. She started walking toward me. I was mortified. Could she not read the sign that was around my neck? Did she not know to leave me alone? But this time she approached me a little differently. She was measured in her movements, as though she didn't want to scare me away. She was still smiling, just not as widely.

She held a piece of paper in her hand. She'd written down her name, *Melissa,* and her phone number. She said that there was a group of widows our age in Concord. She spoke of them as if they were some kind of macabre troupe of acrobats, as though their name should be capitalized: the Widows of Concord. She said that five of them had just met for the first time to help each other through their new realities, their new parts as the abandoned ones. I should join them when they met again, she said. Then she smiled her warm smile and went back to her friend.

I would make six. I stood at the top of that hill and did the probability math. So many young widows in such a small town—Concord's population isn't twenty thousand—seemed highly unlikely. I had announced as much: "That's a statistical impossibility," I'd told Melissa. Then I remembered the previous summer, when I'd called Max and Alex's camp to warn the director that their father was dying. The director said that it wouldn't be a problem. "We're used to it," he said. I was taken aback at the time, but now I understood. Concord had more than its share of fatherless children, gone halfway to rogue.

I kept Melissa's number in my coat pocket. I would pull it out and look at it day after day, making sure it was real. I was terrified that I would lose it, but I was also too scared to call. I'd never met anybody quite like me; why should I now, after I'd become even more of an outlier? I didn't want to find out that the other widows weren't like me after all. Months before, I had called a number I'd seen in the local newspaper, advertising a widows' group, but the woman who picked up the phone had rejected me, saying that the group was for old widows, not young ones. She'd made me feel like a freak. In the middle of such sadness, it's hard to imagine that anyone in the world knows how you feel. And yet somehow there was a small army of women in my little town who knew exactly what I was experiencing, because they were experiencing it, too. Whenever I pulled out that scrap of paper, I felt as though I were holding the last unstruck match in a storm.

It was nearly a week before I got the courage to call Melissa. The paper was nearly worn through by then.

The phone rang. Melissa picked up. She asked me how I was doing. Hardly anybody was brave enough to ask me that anymore, and I didn't know how to answer.

"Okay," I said. "Not okay."

Melissa said that the Widows of Concord were going to have a party soon. She asked if I wanted to come.

"Yes," I said. "Very much. When are you getting together?"

There was a little pause.

"Valentine's Day."

CHAPTER 1

A Stargazer Is Born

I was ten years old when I first really saw the stars. I was mostly a city kid, so I didn't often experience true darkness. The streets of Toronto were my universe. My parents had split up when I was very young, and my brother, sister, and I spent a lot of time on our own, riding subways, exploring alleys. Sometimes we had babysitters barely older than we were. One of them, a boy named Tom, asked my father to take all of us camping.

Camping wasn't my father's idea of a good time. Canadians escape to "cottage country" as often as they can, snaking out of the city in great lines of weekend traffic, aiming for some sacred slice of lake and trees. Dr. David Seager was British, and he often wore a tie on weekends; for him, sleeping in the woods was something that animals did.

But Tom must have made a pretty good case, because the next thing I knew, we were on our way north. We went to a provincial park called Bon Echo, carved out of a small pocket of Ontario, three or four hours from Toronto. Bon Echo includes a string of beautiful lakes, almost black against the green of the trees. There are white beaches and pink granite cliffs—perfect for jumping off into the cool water, after climbing as high as you dare—and thick red beds of pine needles on the forest floor. Bon Echo was the prettiest place I'd ever been.

Maybe it was the absence of city sounds that made it hard for me to sleep. I was in a tent with my siblings. We had set up a little suitcase between us like a nightstand. (As usual, we had been left to our own devices, this time to pack. We had no idea that campers generally don't bring suitcases.) My brother and sister were making the soft noises that sleeping children make.

Jeremy was the oldest and tall for his age. He had only a year on me, but it was a crucial year, and he usually ended up in charge, dictating our daily activities from his great height. Julia was the youngest, beautiful and boisterous with a perpetual light in her eyes. She was everybody's favorite. I occupied the middle in every sense, small and silent. I was the dark one. Jeremy and Julia have blond hair and blue eyes; I have brown hair and hazel eyes. My eyes were also the only ones open that night.

I unzipped the tent's flap and ducked out into the dark. I wandered just far enough away to clear the last of the trees.

That's when I looked up.

My heart stopped.

All these years later, I can still remember that feeling in my chest. It was a moonless night, and there were so many stars— hundreds, perhaps thousands—over my head. I wondered how such beauty could exist, and I wondered, too, why nobody had ever told me about it. I must have been the first person to see the night sky. I must have been the first person in human history who had braved her way outside and looked up. Otherwise the stars would have been something that people talked about, something that children were shown as soon as we could open our eyes. I stood and stared for what felt like hours but was probably seconds, a little girl who understood how to navigate the chaos of a big city and a broken home, but who now had been given her first glimpse of real mystery. I was overwhelmed by what felt like too much light, too much knowledge to take in all at once. I ran back to the tent, curled up beside my sleeping sister, and tried to be just ten years old again, listening to the sweet sound of her breathing.

•

My father lived outside Toronto, in a series of neat and orderly apartments and bungalows. My mother lived in a former rooming house, in what was a battered part of town called the South Annex, with my stepfather, piles of old newspapers, and an army of cats named after literary characters. She was a writer, a poet.

I never became close with children who weren't related to me, so I didn't know how different our family was. When I'm feeling generous I tell myself that we were lucky to live without any of the usual constraints imposed by more conventional upbringings. I learned to believe that freedom is precious however it's given to you, and our almost impossible freedom helped make us who we are today: Jeremy is a nurse; Julia is a harpist; I'm an astrophysicist. But when I reflect on the realities of our young lives, I can hardly believe we survived, especially when I look at my boys at the same age. We were cubs, turned out to run with the bears.

When we first lived in the Annex, we attended a Montessori school far outside town, near the distant house we'd called home before my mother and father separated. I don't know why we stayed in the same school after our move into the city, but our commute was over an hour each way, including trips on two buses and the subway, with long waits at busy stations and platforms in between. Jeremy was maybe eight at the time, which would make me seven and Julia five. After a few weeks of trial runs, we made that trip every day on our own.

Jeremy would save up a pocketful of coins until he had enough to buy a bag of sour-cream-and-onion chips, which we would carefully share. Just the smell of those chips today puts me back on those buses and subways. We filled time by reading newspapers—discarded by adults, or stolen out of the newspaper box after somebody bought one, before the door could slam shut—which I suppose was a positive. We were what modern educators would call "advanced."

One day my sister fell into a muddy puddle at the bus stop that marked the start of our long journey home. After a tearful ride, a woman saw Julia still crying at the subway station and brought her into the women's washroom to clean her up. She took forever, and I shuttled back and forth giving updates to my brother, who stood worried sentry outside. I try to imagine that scenario now—a woman finding three kids under eight on their own, one of them crying and covered in mud. I think today, most of the time, the story ends with a call to the police. In our case, it ended with a stranger putting my five-year-old sister slightly back together before we boarded the subway into the city.

I have memories that left more lasting damage. My stepfather was a monster, the kind of beast who normally lives at the dark heart of a fairy tale. He didn't physically abuse me, but he could be unbelievably cruel, and his mood swings were vicious. I lived in constant fear of setting him off.

He and my mother were both still in bed when we left for school, having scratched together our own breakfast, our own lunch. He didn't work, and my mother's writing career wasn't exactly lucrative, either. My father told me he suspected our entire family survived on his child support payments. When my mother and stepfather had a child together, my half sister, money was so tight that I wondered whether six of us were living off child support meant for three. Julia and I had to share our already cluttered room with the baby. She cried all night for months with colic, and she would wake up at dawn for a long time after, my mother ignoring my pleas to cover our east-facing windows. I was forever getting up to take care of the baby.

When I was nine years old, I decided not to walk with Julia to school one morning. (We had left the Montessori at that point, but our new school was still a mile-long walk away.) She would have been seven. I wanted to walk with one of my few semi-friends and didn't want my little sister tagging along, so I told her to find her own way. Instead of taking the safer, quieter side streets, she took the main roads. At one especially busy corner, she was confronted by an unstable woman who howled in her face and tried to hit her with her bags. Julia froze and screamed for help. It took a long time for anyone to answer her cries. A real estate agent finally surfaced from a nearby office to rescue her. For days after, teachers at our school would ask me what had happened. *"Not sweet Julia!"* They were in shock.

"You are in so much trouble," my stepfather screamed at me when I got home. I can't remember exactly what he said after that, but these are the words I hear when I close my eyes: *You are a bad person. What were you thinking? You are so irresponsible. You are an ungrateful child, and I am furious with you.*

I should have looked after my sister. But I was also nine years old. That night I was the one who woke up crying.

•

We spent weekends with my father, at first in his apartment by the wide-open highways. Those two days felt like vacations from fear. My father took afternoon naps to catch up on the sleep he missed during the workweek, while my brother, sister, and I hung around his apartment playing games, often of our own invention. One afternoon, we went out onto the apartment balcony. He lived on the eighteenth floor. It was the farthest off the ground we had been in our lives, and, pretty naturally I thought, we decided to drop all sorts of objects over the railing to watch them fall. Nothing heavy: a comb, a doll. But gravity is gravity, and everything picks up velocity when it's dropped eighteen stories. We watched our chosen projectiles land and strained to hear their moments of impact, learning a little about acceleration physics and the speed of sound. Then we rode down the elevator, gathered them up, and tried again.

When my father woke up and learned what we were doing, he was apoplectic. We could have hurt somebody, and we weren't supposed to leave the building on our own. I didn't even understand that such rules might exist, let alone that I was expected to follow them. I've since learned that a lot of scientists have mischief in their pasts, and their particular strain of mischief can be a good predictor of their future chosen field. Chemists, for instance, usually experience a period of childhood pyromania. Biologists might get a little too curious about what's inside frogs. Physicists, somewhere along the way, drop objects from heights.

Though he didn't like that experiment, my father was big

on example, a teacher by illustration. His first apartment wasn't built for a family, and we slept in makeshift beds, at least free from the worry that strange-eyed white cats named Rosencrantz and Guildenstern would spray our clothes. One morning I was putting away the pullout I shared with my sister when I accidentally ripped the orange polyester blanket we'd been using. I had been conditioned by my stepfather to expect consequences for such carelessness, and I started crying hysterically.

My father couldn't understand why I was upset: My reaction was so out of scale with the event. Unfortunately, he didn't connect one dot to the other. He had heard us complain about our stepfather, but I think he thought we were the typical children of divorce, angry at our surrogates out of instinct. In the moment, he couldn't see anything beyond his scared little girl, devastated by a tear in a cheap blanket.

I've never forgotten what he did next. He held the blanket on either side of the tear, and then he ripped it in half in front of my bloodshot eyes. He was trying to teach me that there are things that matter and things that don't. But at the time I took a different meaning: Where you are changes everything.

•

As we grew up, I became closer and closer to my father. With him I felt understood. He was a family doctor for years, his bustling practice a cornerstone in the small town of Markham, north of the city. Markham grew into a city all on its own, and my father remained at the center of things. It was a slow build, but he made it, and he moved into a bungalow in the northern suburbs that seemed like paradise to me. My time with him had always felt like a reprieve; now every weekend felt like an escape.

My father saw that I was atypical, that my brain wasn't like other kid brains. He sometimes worried aloud that I was too serious and unsmiling; once, while looking at some photographs, he showed me what he meant—that my eyes were sad and unfocused, as though I were staring at something that no one else could see. He confessed that he'd wondered whether I was developmentally challenged. Decades later I would find a label for my wayward gaze, a diagnosis that puts me on the autism spectrum. For now, and for the rest of my father's life, I was just his daughter who was wired a little differently. I would spend a long time wondering—and agonizing—about my feelings of otherness, but my father gave me the gift of accepting me without explanation.

I can remember he had a friend over for dinner who said that my insides belonged to someone much older, and my father beamed at the idea that my body didn't match my soul. He believed in reincarnation, and he wondered aloud whether we had known each other in a past life and that explained our connection. He was sure that we would find each other again in the future.

By the time I was eleven, books had become my principal means of connecting to the world, and when the subject of reincarnation came up, I did what I usually did and went to the library to read about the prospect of life after death. I came away from my research understanding that death was final, but my father had exposed me to other possibilities. That's what fatherhood was to him: His job was to serve as a tour guide to the marvels of human existence. He decided that I was going to be a doctor like him, and he began grooming me for his particular brand of success. He played me soaring classical music and gave me books that were far beyond my reach. I remember he handed me a George Gamow book called *One, Two, Three . . . Infinity.* I read it, as instructed. It made zero sense to me.

Another book, a thin red paperback called *The Magic of Believing,* did make an impression. My father bought a carton of that book and would hand them out to any taker. It was a historical survey of the power of positive thought. I read it over and over again. My favorite part was a story about a girl named Opal, the daughter of a logger in Oregon, who believed that she was French royalty. Most dismissed her as a lunatic, but by her twenties she had become part of a royal family, albeit in India, where she was spotted by a journalist in a magnificent carriage drawn by a team of horses. That book made me believe in a kind of practical magic: that vision begat planning, which begat opportunity. I could will my way to a better life.

My reality remained resistant to change. When I was twelve, my father enrolled me in a private school: St. Clement's School, for Anglican girls. We were Jewish, in theory if not practice, so I was only a half-fit. It was the only private school that would take me. The entrance exams for all the others had been easy, but the interviews were a different matter. Maybe the schools thought I was too socially unprepared to belong. Looking back, I think the problem was more likely my silence during what was meant to be a conversation. I never knew what to say, so I mostly said nothing.

I entered St. Clement's in the seventh grade. We were forbidden from leaving school property, but I had been walking the streets of Toronto on my own since I was six. There was a bakery across the street that called out to me, and I wasn't going to let some stupid rule stop me from going. A few weeks after my arrival, I crossed the street.

That was the equivalent of arson at a school like St. Clement's, and in a way I did start a fire. Other girls began to question a curriculum that was designed to make us obey. They began cheating in study hall and writing scandalous things on blackboards. (One girl wrote *Jesus Loves You,* which was considered offensive for reasons I never understood.) The principal saw me as the catalyst for the rebellion, probably because I was. She summoned me to her office more than once. "Sara," she began each time, "you are very smart, good-looking, and the other students follow your lead. You could put those traits to better use." Something had changed in me, and I bristled at her judgments. Why should I be the person she thought I should be?

When other parents began forbidding their daughters from talking to me, I realized it was time to change schools. I went back to the public system; within a year or two, I had fully embraced my fate. I fell in with a band of rootless teenagers from schools across the city. Word would go around, and we'd meet up later that night on a random subway platform. None of the kids was my friend, exactly, but two older girls took pity on me and made sure that I came along to parties. They teased me about how I dressed before lending me better clothes, and I trailed behind them like a mascot, trying to figure out how to feel what they felt for each other. (Their teasing was better than the purer tormenting I could suffer at school.) We would pour into sockets of the city like mercury. There was a lot of alcohol. There were a lot of drugs. I might have been my father's daughter, but only on the weekends. During the week I still lived with my mother and stepfather, and on those five nights I tried to stay far out of sight.

•

In the late winter and spring of 1987, when I was fifteen years old, a new star appeared in the southern sky. A blue supergiant named Sanduleak -69° 202 had exploded in the Large Magellanic Cloud, which is a small satellite galaxy next to the Milky Way. It was the closest supernova in nearly four hundred years, the first opportunity for modern astronomers to witness firsthand the death of one star and the birth of another. It was 168,000 light-years from Earth, but you didn't need a telescope to see it: From its February discovery through the peak of its brightness in May, the last of its light hung in the sky. It was only after the light disappeared that astronomers were able to confirm that Sanduleak -69° 202 was the lost star.

One Sunday afternoon, I was supposed to go skating with some girls from my school. I bailed and went instead to a presentation about the new supernova at the University of Toronto. Among a panel of men in suits, one man was conspicuously in jeans. He turned out to be the astronomer who had discovered Supernova 1987A and its halo-like light. Two thousand people, seated in rows radiating out from the stage, listened to him speak. I sat enthralled in the pin-drop quiet, ravished by an amazing tale of discovery. The sense of wonder that had overwhelmed me in Bon Echo was reignited. All it took was the self-destruction of a star.

Later that summer I turned sixteen. I stopped running with my crowd of teenagers. We were on a ferry to the Toronto Islands, trying to kill another boring, endless night, when I saw the lights of a boat going the other way and realized that I wanted to be on it, not the one I was on. I got a job working a carnival game at the Canadian National Exhibition, the game with the impossible-to-catch plastic fish. After three weeks in

the crowds and the heat, I'd made the vast sum of $400. I spent
every penny of it on a four-inch reflecting telescope.

I kept the telescope at my father's place. I spent chilly weekends the following winter standing in a sprawling parking lot,
looking up at the stars. My father often stood shivering beside
me, our breath turning into a single cloud in the cold.

I can remember with perfect clarity the night we found Jupiter.

•

Back on Earth, my father decided to embark on a second career: He began offering hair transplants. Despite his success in
internal medicine, he enjoyed the feeling of starting again,
pouring the foundation for another slow build. I thought there
was something bittersweet about his new work. He wasn't
saving anybody's life by giving them hair again, but his new
patients became some of his most faithful admirers. They had
endured years of stress and insecurity, the pain of an inevitable, undesired conclusion, and here was a man who promised
to restore everything that they had lost along with their hair.

Early hair restoration was barbaric. Desperate men sub-
mitted to having hundreds of plugs cut out of sections of their
scalps. The surgeries could leave them more damaged and vul-
nerable than they were before, the cure worse than the disease.
Something called "shotgun scarring" was a common side effect. My father wanted better, and he was obsessive about im-
proving his technique, making his trademark thousands of
more realistic-looking, single-follicle grafts. He experimented
with every promised advance—he was among the first to use
lasers before he rejected them for scorching what they were
meant to sow—and he never seemed satisfied with even the
best labors of his practiced hands. A reasonable-seeming hair-

line doesn't seem like it should be the most elusive goal in the world, but nothing defies mimicry like nature, and my father's devotion to his practice and patients had its impact on me. It was the most accidental yet meaningful of his illustrations. There was something inspiring about his refusal to accept the present as a forever reality.

After brief stops at other schools, I finally landed at Jarvis Collegiate Institute, a public high school near the heart of the city with an excellent reputation for math and science. It was diverse in every sense, filled with immigrants from all over the world, a dizzying collection of sophisticates and loiterers, geniuses and stragglers. Jarvis Collegiate was the perfect school in which to be a loner. There was no pressure to belong, because nobody there could agree on what it meant to be cool. I didn't feel the relief of connection. I felt the relief that comes with not having to worry about finding connection anymore.

I was walking to school one day, by myself as usual—cutting across the divided campus of the University of Toronto, the old half made of stone, the new half made of glass—when I saw a sign for a school-wide open house that weekend. On Saturday, I returned and found the elevator in the tallest building on campus and pressed the button for one of the upper floors. I stepped out into the astronomy department. There was a table staffed by professors and students handing out small piles of paper, and it struck me in an instant: Astronomy could be more than a passion; it could be a career. I made up my mind to commit myself to my schoolwork. Good grades would get me into university, which would allow me to look at the stars for the rest of my life. Magic.

Most subjects proved easy for me—with the notable exception, initially, of physics. It was hard for me to apply its equations to the real world; life seemed more random and chaotic than that. My life did, at least. Then one day my physics teacher gave us each a small coiled spring. On the other side of the classroom, he set up a board with a hole cut into it. The object of the exercise was for us to calculate the spring's force constant, and to use Hooke's law and the equations of motion to find the perfect angle at which to launch the spring across the room and through the hole.

One by one, we made our attempts. Maybe a third of the students found the mark. (I have my suspicions about how many of them had followed Hooke's law and how many were just lucky.) I did the math, checking and double-checking it before it was finally my turn. I angled my spring and fired. My mouth fell open while I watched the spring arc perfectly across the room, straight through the hole.

•

At the start of my last year of high school, I was surprised to be handed three envelopes along with my class schedule. I opened the first to find a letter that said I'd earned the top marks in my entire grade the previous year, finishing first out of three hundred or so students. The other two were subject awards. I didn't even know our school gave out academic awards—I'd never received one, and I'd always skipped the assembly when they were presented. A couple of days later we gathered in the school auditorium. I was in the school band and we played before the awards were presented. Each time my name was called, I had to put down my flute and walk across the stage. I felt awkward, maybe even embarrassed, as a small pile of certificates soon joined my sheet music.

One of my former party friends, now a stranger to me, came up to me afterward in the hall.

"I didn't know you were so smart," he said. I can still hear the way he said it, with a weird mix of anger and smugness and confusion. He had wanted to be my boyfriend at some point, but I didn't feel that way about him. Maybe he saw his chance to reject me back.

"Neither did I."

I suppose I should have been happy or proud of my achievement, but I wasn't especially. I looked at it with logic: In the subjects in which I had achieved the highest grades, I won the awards. That made sense. What made less sense to me was that I had made the highest grades the first time I tried to earn them. I hadn't been relentless or anything like single-minded. I had simply decided to work harder. The math didn't add up. It should have been tougher to be the best.

My father was happier than I was, until I told him one more time that I didn't want to be a doctor. Since the open house, I had insisted to him that I was going to be an astronomer. He gave me a harsh lecture during my next visit. It was one of our few weekends together that felt too long rather than too short.

"You have to get a job and support yourself," he said. "And. Not. Rely. On. Any. Man." My father's resistance to my ambitions struck me as ironic. A psychic had once told him that he'd be a household name, and by the early 1990s he had earned an unlikely fame for his own unconventional path: The Seager Hair Transplant Centre and its ubiquitous billboards still celebrate him more than a decade after his death. He credited no small part of his achievements to *The Magic of Believing*. But when it came to his daughter's future, he wasn't quite so willing to challenge the Fates.

Nobody made a success of themselves by trading in abstraction, he scolded. "The world wants evidence," he said, nearly shouting. "The world wants proof." I heard him, but I didn't listen. Jupiter had already made its greater case.

•

There's a famous play, *Equus,* about a troubled boy with a blinding love of horses. The boy sees a psychiatrist named Martin Dysart, who tries to understand him by trying to understand his love. Dysart is confounded by it:

> A *child is born into a world of phenomena all equal in their power to enslave. It sniffs—it sucks—it strokes its eyes over the whole uncountable range. Suddenly one strikes. Why? Moments snap together like magnets, forging a chain of shackles. Why? I can trace them. I can even, with time, pull them apart again. But why at the start they were ever magnetized at all—just those particular moments of experience and no others—I don't know.*

I can trace my love, too. Why stars instead of horses, or boys, or hockey? I don't know. I don't know. Maybe it's because the stars are the antithesis of darkness, of abusive stepfathers and imperiled little sisters. Stars are light. Stars are possibility. They are the places where science and magic meet, windows to worlds greater than my own. Stars gave me the hope that I might one day find the right answers.

But there's more to my love than that. When I think of the stars I feel an almost physical pull. I don't just want to look at them. I want to know them, every last one of them, a star for every grain of sand on Earth. I want to bask in the hundreds of millions of suns that shine in the thousands of billions of skies in our galaxy alone. Stars represent more than possibility to me; they are probability. On Earth the odds could seem stacked against me—but where you are changes everything. Each star was, and still is, another chance for me to find myself somewhere else. Somewhere new.

CHAPTER 2

A Change of Course

There were thousands of miles between us and the top of the world, and I hadn't seen a single inch of them. Every step ahead of us would be pure discovery. I felt a charge run through my entire body: the electric thrill of the unknown. Up there, I knew nothing.

According to the boundaries on our still-crisp map, we were in northern Saskatchewan. Now those lines seemed meaningless to me, a futile attempt to impose order in the absence of anything human-scale. Saskatchewan is a giant rectangle on paper, but we were in a place that defied such conventional geometry. There were no landmarks, no cross-roads, none of the usual signs or sharp corners that we use to find our way. There were only rocks, trees, and rivers, stretched out in a tangle as infinite-seeming as time.

It was June 1994. I'd just finished my bachelor's degree in math and physics at the University of Toronto. The past two summers, I'd interned at the David Dunlap Observatory not far outside the city, dividing my time between observing and classifying variable stars—stars whose brightness varies—and reading the leather-bound astronomy books that I pulled from the ladder-tall shelves. I'd also been drawn more deeply into the wilderness, taking canoe trips under the canopy of lights I'd first seen at Bon Echo. I decided to take a long break before I waded into my graduate work at Harvard and devote it to the trip of a lifetime: two months in a canoe in Canada's remote northern reaches, beyond the last of the trees.

The focus and discipline of university had erased the last vestiges of my vagabond adolescence, but I remained restless, given to fits of mental and physical wandering. I was never satisfied with the world in front of me. There always had to be more. Once again a book changed the course of my life: *Sleeping Island*, written by a man who left his schoolteacher life in Boston one summer to explore the great Barren Lands by canoe. It set me dreaming about an epic traverse. I spent my last undergrad winter in the library, poring over maps and accounts of century-old expeditions written by the low light of lanterns. Even in sepia, the Arctic was an otherworldly landscape, made up of nearly as much water as earth. In the north in summer, under the glare of a midnight sun, there could seem to be more lakes than stars.

I joined the Wilderness Canoe Association in Toronto to prepare for my own journey. One weekend I needed a ride to a backcountry ski trip (organized while we waited for the rivers to thaw), and a club member named Mike Wevrick offered me a lift. When I arrived—late—for our rendezvous, I found him in his beat-up car reading a slightly less-worn paperback. Between his book, his beard, and his mop of ginger hair, I couldn't see much of his face. His eyes were his only defining feature, the same blue as the winter sky.

We spent five hours in the car together, bound for Killarney Provincial Park near Sudbury, Ontario, where we skied with a group through the forest, the trees clinging like mountain goats to the steep terrain. Mike said that he was impressed with my skiing. I wasn't impressed with his, especially when he decided to end the day early and grab a donut from a nearby Tim Hortons. I wanted to ski until dark.

Mike called me again and again after our trip, trying to convince me to go on another adventure with him. He probably called me twice a week for the better part of a month. I rejected him exactly as often. I thought I understood what he saw in me—I really was a pretty good skier—and maybe a little of what he saw in us. We had found plenty to talk about on our long car ride, and we both loved the outdoors. That was it, really. Did that warrant our spending more time together? The truth was, the highest register on my human-companionship spectrum at the time was *Tolerate,* and I didn't bring new people into my life unless they gave me a really good reason.

I had felt a tiny spark with Mike, but nothing like the lightning strikes you see in movies. Was a tiny spark a good enough reason to let him in? I didn't think so. Besides, I'd be leaving Toronto at the end of the summer. Harvard's Department of Astronomy had accepted me into its graduate program the day we'd gone skiing at Killarney. There was no point in starting something that would end before it had a chance to begin.

"Do you want to go skiing again?"

"No, thank you."

"Do you want to go hiking in the White Mountains?"

"No. But don't take it personally. I'm leaving in September and I'll be closer to the White Mountains anyway."

Then one day in March, Mike called with word that the ice had broken on the Humber River: "Do you want to jump in a canoe?" The Humber ran through the city and wasn't especially picturesque, but I loved paddling more than anything.

Mike finally heard me say yes, even if it was the water I wanted, not him.

The following weekend saw us pushing into a set of artificial rapids—the spill from a dam—and we began rehearsing white-water maneuvers. We were out of practice and sync, and within a few minutes we had capsized Mike's canoe, a battered Old Town Tripper. I'd worn a wetsuit, but I was still cold and wet and not all that happy with Mike. It wasn't until later, back at his house for a warm-up, that I fully realized he'd shaved off his beard and trimmed his hair into a crew cut. He looked better cleaned up. He'd also stripped off his wetsuit down to the pair of tiny shorts he'd worn underneath. His muscles rippled like the water we had paddled. *Wow,* I thought. *He's cute.* I wondered whether starting something wasn't such a bad idea.

We paddled together a lot that spring. We began to click in inexpressible ways and started our reasonable facsimile of dating. Even though he was my version of a boyfriend, I preferred to call him my canoe partner. I was thrilled to find someone with whom I could share a boat. Every one of our dates included time on the water, and we built a quiet understanding. Our talks meandered between sets of rapids like the river itself, and our paddling filled the silences. Mike was an editor who worked with words the way I studied light. We both spent a lot of time inside our own heads, trying to bend elusive things into shape. We found that we could be alone together.

I told Mike about my ambitious travel plans for that summer. Like most of my plans, they didn't include company. But I knew that he would see what I saw in my dreams: wild rivers, untouched forests, the abandoned Old North Trail tapering into the blankest possible canvas. My imagination had become a storybook, the title of each chapter the name of another lake: Kasba, Ennadai, Angikuni, Nowleye, Casimir, Mallet. Their Native and Inuit names sounded so exotic to my ears. Over the course of the next few weeks, Mike hinted that he wanted to join me in exploring them. The more I thought about it, the more I had to make an important concession to reality: It was a little crazy for me to think I could tackle the trip on my own. Mike would make an ideal partner, in more ways than one. I said okay. *Why don't you join me?*

One night, we took a break from preparing for our adventure to go for a walk in a heavy rain. Mike held a black umbrella over both of us. On that pitch-black night, over the sound of the rain pouring off the roof that he'd made above our heads, Mike made up his mind to speak. "I've never been this comfortable with someone before," he said. I don't remember if I agreed out loud, but inside I nodded. I was still learning how to navigate my widening emotions, and I marveled at myself for a few minutes. I was charged with excitement, eager about everything that was ahead of us, whatever everything might be. I had never felt that way before. It was like finding out that your heart could do something new.

•

The itch to go north had been mine but the canoe was Mike's, his Old Town Tripper. After launching into the relative calm of a river, we paddled into our first of those alien lakes, an inland sea called Wollaston. Looking across its numbing expanse, I wondered how I had ever thought that I might make such a trip alone.

After crossing Wollaston, we spent the first two weeks of our trip mostly river-bound. We shot terrifying, thumping rapids. It often felt as though we weren't choosing where to go; the rivers almost always made our decisions for us. They were swollen with melt, and we were sucked into rapids that were well beyond my expertise. You have to be careful traveling in a single boat in such a lonely region. You can scream as loud as you want and no one will hear you, and if you lose your canoe, the chances are good that you'll be lost with it.

There were often rapids too big for us to run. I liked watching Mike in those moments, surveying the steam of the river ahead of us. I admired the calmness of his decision-making. We'd pull up to the bank and begin hauling our canoe and hundreds of pounds of gear on our backs. It felt as though we were traveling in the truest sense. We earned the ground that we covered. Northern Saskatchewan overwhelmed us with its rough-hewn beauty, its eskers covered in white spruce and the bleached remains of ancient travelers: fire rings and tin cans, old boots and caribou bones almost silver in the sun. We never saw another living soul, unless you believe that blackflies have souls, in which case we saw millions of them.

A lot of the ground had been burnt over, and there was a haze and the smell of smoke in the air. We had no way of knowing whether the low Arctic's annual bloom of forest fires was getting started or burning itself out, whether we were walking into or out of danger. We knew that the fires were out there, but they were invisible to us. We could judge them only by the mysterious patterns of their smoke.

Nearing the end of our journey upstream, we were paddling through a narrows, maybe 100 feet across, when we drifted into smoke of a different character. It was almost solid, as opaque as a wall. We got out of our canoe and climbed an esker to get a better view of what lay ahead. For the first time we could see the actual fires, their flames reaching for the sky like fountains.

"I don't think we're in any real danger," Mike said. He was optimistic like that, often naively, I thought. In his wishful thinking, he was able to ignore the plainest unhappy evidence. I was more led by facts and their simple calculus. If I looked up at black skies and said it was going to rain, Mike would counter that it might not. My analysis usually proved correct; unfortunately, this time was no different. The instant he finished his hopeful thought, a nearby stand of trees went up in flames. The inferno arrived so suddenly, in a blaze of orange accompanied by oily gray smoke, it felt as though a bomb had gone off.

Now the fire was loud enough to hear, roaring with white noise, like the rapids I wished we were running to escape it. I was paralyzed with fear. Those flames could leap across the narrows and consume us in an instant, and I struggled to convince my legs to carry me back to the canoe. Mike stared at me with a look that's still hard for me to describe: equal parts worry and resolve. We both thought we were doomed. We both tried to imagine we weren't.

We retreated back to the top of the sandy, relatively treeless esker and hunkered down for either the shortest or longest night of our lives. I peeked outside the tent in the twilight of the short northern night, hoping against hope that the smoke had cleared in the semidarkness. It was so thick I could barely breathe.

I decided that we were going to suffocate in our sleeping bags, our bones joining the carcasses of all those fallen caribou. Somehow I fell asleep, and I had the most vivid dream, that we woke up to a few small, smoldering fires scattered across the landscape. When I woke up, the winds had changed. The smoke had thinned, and in a couple of hours the fire had been swallowed by the ground. Mike and I had never packed up camp so quickly, and we began to struggle up the last of the narrows. The water was maddeningly low. We had to wade, dragging the canoe over the rocks, for what felt like interminable passages. At last we broke into the cool, wide-open waters of Kasba Lake. We found an unlikely safety in its expanse.

That night on the esker moved me deeply. I'd studied physics, and physics is a science founded on the rules of logic and law. I knew then and I know now that the weather is governed by geographic and atmospheric forces, that no power higher than a fortuitous change in the wind conspired to save us that morning. It was still humbling to witness how much of our lives can depend on forces beyond our control.

We stopped at a remote fishing lodge on Kasba Lake, where we had shipped ahead our resupply. It was the only permanent shelter for hundreds of miles. We ate our weight in home-cooked lake trout and spent the night in a cabin—reveling in the bed, resenting the roof—and then continued our trip north. We paddled beyond the last of the boreal forest, into the treeless tundra of what is now called Nunavut. It was a new world. We began seeing herds of live caribou, not just their skeletons. We stumbled upon Inuit graves. We caught giant fish and cooked them on stone beaches. We spent week after week navigating our way through difficult stretches of river and portaging across boulder fields, plunging our canoe into lakes in moments of ecstatic release.

Mike and I had found a routine all our own; we didn't have to make room for anybody or anything other than each other. We had both stopped wearing watches. The sun was our clock. We ate when we were hungry, which was most of the time. We slept when we were tired, which was the rest of the time. The calendar that we had nearly forgotten existed finally forced us to loop around and head back south, returning from what might have been the surface of the moon to the subtle signs of humanity and the unfamiliar shelter of trees.

"I've decided I really like trees," Mike said.

"I've decided I really like trees, too," I said.

We ended up back at the lodge on Kasba Lake and had our first conversations in weeks with people other than each other, strange faces the starkest reminder that the universe was larger than our canoe. Mike and I had lost a day along the way, and we arrived just in time for the season's last flight out. I didn't feel relief. I was devastated to join the geese on their airborne migrations. I felt sorry that we weren't forced to winter over, as though this time we had been on the wrong side of a close call. I had spent sixty days living my perfect life: alone except for one dreamy companion, together visiting places that no one had seen, escorted by just enough fear to feel a constant low crackle.

I hadn't been struck by lightning. But I had fallen in more than one kind of love.

CHAPTER 3

Two Moons

The trip changed me. I emerged at the end of it not just a different person; I felt like a visitor to my former world. I had lost the calluses that protect us from its constant bumps and bruises. Former inconveniences—sitting in traffic, hearing a phone ring, suffering through inane chatter—felt like nightmares. Only a little pollution made my lungs hurt; spending time in closed-in places made the rest of me ache even worse. Harvard has a bucolic campus in the fall, an immaculate collection of red bricks and turning trees, and Cambridge isn't Los Angeles at rush hour. I still struggled under a psychic weight, as though the sky were a ceiling that was too low.

I lasted about two months in a graduate student dorm on campus before I fled to Shirley, Massachusetts, an old Shaker village in Middlesex County. I found a nineteenth-century carriage house that had been converted, not all that well, into a residence. It was heaven for me. There was a Christmas tree farm across the road, and I became close to Beth and Will, the couple who owned it. There was a pond for swimming or a peaceful paddle. When it snowed, I could go cross-country skiing for hours right from my door. There was also a pretty river nearby, the Squannacook, with a gorgeous stretch of rapids I visited whenever big rains came and the water was high enough to run.

Mike and I moved in together before winter fell. Like his joining me on my canoe trip, our living together was hardly planned. We let ourselves be guided by the river of circumstance. I had first raised the possibility on our drive back from Wollaston Lake. I reached out and grabbed his arm. "Mike, move to Boston with me," I said. His blue eyes went bluer for the water that began to fill them, but he didn't reply. His wishful thinking abandoned him whenever he wondered how we might find ourselves in the same place for any length of time.

I missed him after I'd moved, and I clocked that new feeling, too: *I want someone else with me.* I wrote him letters. We talked on the phone. One day, a month after our return, he told me that he'd been laid off from his job in Toronto. Taking the summer off had been a great way to prove how inessential he was to the operation. He told me that he was thinking of moving back in with his mother in Ottawa; that didn't make sense to me. He was thirty years old. Why shouldn't we try living together? Mike couldn't come up with a reason, and he moved down to Shirley with his canoe and tiny shorts. He soon found work as a freelance editor, poring over science and math textbooks in the pale light of the carriage house. His job was to find mistakes. I went to school, where my job was to risk making them.

•

The 1990s were an era of historic discovery in astrophysics. Our tools were catching up to our ambitions. More powerful computers and satellites allowed us to make calculations and take measurements that we couldn't have contemplated only a decade before. In astronomy, there was always something new for us to do.

In 1995, during my second year at Harvard, I was a little adrift, still searching for a specific course of study. NASA was building a new satellite; it would eventually be called the Wilkinson Microwave Anisotropy Probe, or WMAP. It was designed to observe cosmic microwave background radiation, that oldest, Big Bang light. My research adviser, a young Bulgarian astrophysicist named Dimitar Sasselov, suggested that I might find an academic home in the shared effort. He was correct. Together we would divine the origins of the universe. I caught the first glimpse of my calling in its ancient glow.

About 380,000 years after the Big Bang, the universe was still a white-hot fog, billowing at its limits like the farthest reaches of an explosion. It was too hot for the formation of atoms, so protons and electrons drifted in the haze, angry and anchorless. The universe cooled as it continued to expand. Eventually it became cool enough for the protons and electrons to start combining with each other when they collided, creating the first hydrogen atoms. That hydrogen later formed the hearts of stars.

The universe expanded so quickly that some of the electrons failed to find a proton in the chaos. What we think of as "empty space" isn't empty; it contains not only those lonely, leftover electrons but also the energy they once scattered, detectable to us as radiation. (Astronauts see that same radiation whenever they try to sleep in orbit; every minute or so, lights flash on the other side of their closed eyelids.) Today that lingering energy is faint, but there remain slight differences in temperature across space. With WMAP, the new satellite, astronomers would soon be able to map those temperature differences and use the variations to trace the origins of galaxies, the way arson investigators read burn patterns to find the source of ignition. That would allow astronomers to determine when and how galaxies were formed. Their seeds were just waiting for us to find them, frozen in time and space.

In the 1960s, astrophysicists had done their best to calculate the probable rates of cooling, which helped put a rough time stamp on the birth of galaxies. My job, three decades later, was to use modern computers to verify their work. The temperature measurements taken by WMAP were only useful if we interpreted them correctly. I was there to improve our precision.

With code that I'd written from scratch, I ended up finding a tiny discrepancy between the estimates of the 1960s and the actual order of things—an almost imperceptible difference in the accepted timing of when all those protons and electrons finished combining to form hydrogen. It was, at its essence, a microscopic gap between the best guess and the measured reality, but when you're working on such enormous scales, the smallest mistakes can be amplified into massive miscalculations. I had made a small but significant correction to a milestone of the universe.

Or rather—science had, as it almost always does, corrected itself. It's a discipline of constant catching up. My contribution didn't make me a prodigy or someone to watch overnight; I was a kid from Harvard who had made an important, but not unexpected, adjustment. Nothing changed about my existence except that I understood a little better how we come to know things. We make progress the way Mike and I covered our hard northern ground: in the long and steady accumulation of increments.

It wasn't until years later, after WMAP finished its scan of the sky in 2010, that our efforts in the 1990s came to their final fruition. I still marvel at what we now know. First, the universe underwent an extremely rapid growth rate in a tiny fraction of the first trillionth of a second of its existence. That's why someone like me talks about the Big Bang rather than the Big Bang Theory. Second, the universe is about 13.75 billion years old and still growing. We were born in a flame that has never gone out.

•

Though the move to the carriage house and living with Mike helped, I still sometimes suffered at Harvard, trapped in its particular isolations. Every university department has its own elaborate systems of belonging, and I was aware that I wasn't part of any of them. I had struggled to fit in during undergrad, despite its fuller classrooms and forced collegiality, but at least I was from Toronto. I had the comforts of home and those well-worn routines that allow you to mistake familiar bus drivers and clerks for friends. Graduate school made it harder for me to imagine my way out of my solitude. I watched my fellow students the way biologists might observe a family of apes. They formed bonds with each other, but I couldn't figure out how or when.

I also struggled to connect to my work. I loved the stars as much as I always had, but studying astrophysics could make them feel farther out of reach than they already were. Our days were exercises in abstraction and tedium, light reduced to algorithms. It was as though I had decided to study architecture because I loved LEGO, and then found myself in class after class dedicated to the vagaries of the building code. As validating as it was to contribute to our understanding of the universe, my day-to-day just wasn't what I'd thought it would be.

I wasn't unusual in feeling this way. Astrophysics moves at the literal speed of light, and most of us students weren't equipped to know what might prove meaningful three or four or five years later. It was one thing to accept that progress was piecemeal, but our successes were so small in the scheme of things that it was hard for me to find a lasting sense of purpose in any one project. During that second year of school, I thought seriously of quitting.

I daydreamed about going to veterinary college instead. Saving sick or broken animals seemed a lot more practical than exploring the theoretical limits of our universe: *An animal was going to die; but now, because of my knowledge and care, it will live.* Or I could keep trying to determine what happened during the first trillionth of a second after the Big Bang and the 13.75 billion years since. I called my father for solace. "Oh, honey," he said, "that's normal for grad students." But he remained ever the opportunist. "You know," he said as casually as he could, "if you want to change your mind and go to medical school, I'll pay for it." Perhaps strangely, the idea that he would need to reinvest in me and my education helped me decide: It was too late to turn back now. I had made my choice, and that was it. I felt almost bound by the physics I studied. Momentum was a powerful force.

Another force, as strong as it had been that fiery night on the esker, was luck. Right around the time I was completing my work on the early evolution of the universe, Swiss astronomers found the first widely recognized exoplanet.

The greatest discovery astronomers could possibly make is that we're not alone. Humanity has searched the heavens for a reflection of ourselves for centuries; to see someone or something else, inhabiting another Earth—that's the dream. For this reason, among others, the colossal 51 Pegasi b was a major find. It was the first new world found orbiting a sun-like star since Pluto's planetary tenure. It was a tiny crack in the largest possible door.

The Swiss astronomers who discovered 51 Pegasi b didn't really "see" their precious find. The ideal, of course, would be for us to be able to see with our own eyes evidence of other life in the universe. But the best pictures of distant celestial bodies still look like the earliest video games. A handful of pixels, frozen in different shades of white, might represent an entire star system.

That's because they're so far away. Driving the speed limit to Alpha Centauri, the nearest star grouping after our sun, would take us about fifty million years, while our fastest current spacecraft would make the trip in a comparatively brisk seventy thousand years or so. A journey across the Milky Way in that same rocket would take about 1.7 billion years. And the Milky Way is one of hundreds of billions of galaxies. Outside its homey confines, there is another galaxy, and another one, and another one. The universe isn't endless, but it's as close to endless as we can imagine.

Until we can see something with our eyes, we have to find it in our lines of code—one of astronomy's other ways of seeing. We might not be able to gaze upon a particular exoplanet, the lights of alien cities stretching across its surface like spiderwebs. But we can surmise that an object in space exists because of its impact on our numbers. Using a complex mathematical method dubbed "radial velocity," based on the Doppler shift, those pioneering Swiss astronomers witnessed 51 Pegasi b's gravitational effect on its star and deduced that it must be there. It was like believing in Bigfoot because you found his footprints.

As with those infamous plaster casts of giant feet, radial

velocity left plenty of room for skeptics to dismiss the Swiss claims. So did 51 Pegasi b's nonsensical orbit—its "year" raced by in only four days. Similar reports of exoplanet discovery had long been debunked. In 1963, a Dutch astronomer named Peter van de Kamp, then working at Swarthmore College in Pennsylvania, announced that he had found an exoplanet. Like the Swiss, he had deduced his planet's existence by noticing an apparent "tug" on Barnard's Star, thirty-six trillion miles away. Years later, that tug, and the star's seeming shifts in location because of it, was found to be the result of changes to van de Kamp's telescope and its photographic plates. The smallest mistake had been amplified into a massive miscalculation.

Now, with 51 Pegasi b, two rival camps emerged. Three, actually. There were those who accepted the discovery, including my adviser, Dimitar. He was only in his mid-thirties, new to the faculty at Harvard and young enough to remain given to belief. Then there was the anti-camp, led by an aggressive astronomer named David Black. Some in this camp argued that the Swiss had found not a planet's effect on its host star, but rather a new kind of stellar pulsation—that the star wasn't being tugged but was expanding and contracting, the way stars older than the sun do. Or, as Black believed, the Swiss were seeing the effect of one star on another. Perhaps 51 Pegasi b was a brown dwarf or a small star, not a planet. The third camp was our community's version of agnostics. Maybe the Swiss had found an exoplanet; maybe they hadn't. Because 51 Pegasi b was so far away, it didn't matter and never would.

Dimitar suggested that I turn my attention, and my budding knack for finding unseen things, toward exoplanets and the embryonic effort to understand their possibilities. I liked the idea. I would be looking for something tangible, something singular and concrete, and turning my compulsion for wandering into practical astrophysical research. There was no greater unknown than the universe. And besides, what did I have to lose? I remember looking out the window of the carriage house and thinking: *Why not?*

There were in fact plenty of reasons why not. Sitting here a quarter century later, it's hard both to remember and to believe how controversial exoplanets were at the time. Logic dictated that they had to exist. The sun couldn't be the only star that had accumulated planets. But proof of their existence, never mind their potential inhabitants, remained as out of reach as they were. In hindsight, it's amazing that Dimitar assigned a graduate student something so risky, with such a small chance of payoff. Neither of us knew enough to be scared.

Dimitar handed me the rudimentary computer code that had been used to study the effects of stars on each other. How does one star heat another, and what does one star do to the other's stellar atmosphere? He wanted me to rewrite the code so that it could be used to study the effects of a star on exoplanets in close orbits. We wouldn't find life like us on the giant, irradiated planets that orbited so near their stars, but these so-called Hot Jupiters were still worth knowing. I had a hunch that there were lessons hidden in their atmospheres, especially. Perhaps their skies would help us one day know whether we were looking at another Venus or Mars, or another Earth.

I would be studying something a large percentage of the community thought didn't exist or didn't care to know about, and doing so in a way that made the impossible seem even less likely—like trying to prove that Bigfoot exists not by finding him or even his footprints, but by seeing his breath. How could we see the thin envelope of alien atmospheres when we couldn't even find the worlds themselves? I was at a conference when a student from another school approached me in a whisper, asking if I wanted to talk to his adviser. He could explain to me why the Swiss signal couldn't possibly be a planet. A professor from Harvard, my own school, radiated a similar skepticism: We would never be able to detect many exoplanets, let alone their atmospheres. I remember feeling as though people were trying to rescue me from a cult.

In an accidental way, my months in the wilderness had inoculated me against such criticism, however well-intended. My power to focus had been developed like my shoulders. The raw challenge of exploration was so appealing that I didn't really care what anybody else thought about my pursuits. And for all the doubts I had felt since arriving at Harvard, for all the times I felt propelled by circumstance, I was far from passive after I had made up my mind about something. I hadn't seen any reason to date Mike, and later I heard myself asking him to bring his boat to Boston. I'd been ambivalent about studying astrophysics, and now I was determined to help understand brand-new worlds. Once committed to a destination, I was going to get where I was going.

By the time I had rewritten the computer code, in 1999, a couple dozen more exoplanets had been discovered, all by the "star-tugging" method. They were all like 51 Pegasi b, massive with short orbits. (The more massive the planet, and the closer it is to its star, the more dramatic and clear its gravitational effects.) There were still plenty of skeptics, but the evidence against them was beginning to mount, and some betrayed hints of a countering curiosity. Amid these shifting currents, I prepared to defend my thesis. I would argue that one day we would do more than find exoplanets. We would be able to see the light of their skies.

I booked the room in Harvard's Phillips Auditorium used for PhD-defense seminars. It had more than a hundred seats, split between the main floor and a balcony, the walls of each filled with bookshelves. I would be using an overhead projector to show my work, and I flipped through my raft of illustrative plastic sheets with Dimitar one last time. In the middle of my rehearsal, I stopped and worried aloud that the people at the back of the room wouldn't be able to see the finer points of my graphics.

Dimitar laughed. "Sara, there won't be anyone at the back of the room."

On the day of my presentation, I arrived early and set up. A few people came in. Then a few more, and a few more. The room was soon packed full, standing room only.

Exoplanets were for real.

•

All the while, Mike and I continued our simple shared existence. I would go to school and get lost in space and code. I would come home to boats and piles of paper. Mike grounded me, unwound me. He gave me some of the happiest days of my life, long stretches of brain peace. We never raised our voices at each other; I think back on that time and remember the quiet. We spent our springs and summers in the near-silence of our canoe, making several more long trips north, and at home we lived together the way we paddled: It wasn't always easy, because in some ways we remained two people who were built to be alone, but we worked to find a natural rhythm. We spent weeks at our respective work and weekends at our shared kind of play. We hiked and cross-country skied and paddled our way across stretches of Massachusetts, New Hampshire, Vermont. There was still something almost accidental about our connection, and the increasing seriousness of it all sometimes daunted us both. But our pauses never became breaks. Within a year, we had really started to set up camp.

First we adopted a gray-striped tabby that I named Minnie May, after the sick girl saved by Anne in *Anne of Green Gables*. (I had subconsciously inherited my mother's penchant for naming animals after literary characters.) Mike had been resistant to pets—he was so afraid of having anything like dependents, the thought had given him chest pains—but then we agreed that Minnie May needed a friend, and we took in another kitten who turned into a big fat cat named Molly. She became Mike's cat, curled up near him whenever he did his work. Later came a feral black cat named Cecilia that I never managed to tame.

Cecilia's spectral nature surprised me. I have always felt connected to animals. I think it's because, unlike people, they are easy for me to read. Their needs are finite, physical as often as emotional, and I know how to meet them. Animals don't get puzzled or angry when I say the wrong thing. They have short memories. They don't cast judgments or see weakness in difference. They don't take my energy and concentration; they give those precious things to me. Animals are blind to everything but love. Animals forgive.

A dog seemed like the next logical addition to our kind-of family. Soon Mike and I were back at the shelter, adopting what we thought was a Labrador cross. We named her Kira—well, Mike named her Kira, after one of Ayn Rand's spirited protagonists—and she turned out to be some mix of terrier and pit bull. She was strong and willful, with a giant ridge on her head that worked like a flying buttress for her massive jaws. Mike had second thoughts about keeping her, but I thought she was beautiful, and he began to see what I saw in her. He started taking Kira in his canoe, and she became the best kind of ballast.

I thought of Mike and me as celestial bodies, distinct from each other but tied together by the universe's invisible forces. We were like the two moons of Mars: Phobos and Deimos follow different trajectories, but they act in strange, satisfying concert, like the twin sons of Ares and Aphrodite for which they are named. The two moons are relatively small, so they remained hidden from us until Asaph Hall, an American astronomer, found them in 1877. Hall must not have been filled with the warmest thoughts when he named his discoveries after the personifications of horror and terror, but there was an overarching logic to his nomenclature. Mike and I somehow named pets after characters from Ayn Rand and *Anne of Green Gables*.

At least we both loved to read. Mike was a libertarian who believed in the supremacy of the individual; I believed in the supremacy of the universe. Mike thought deeply about philosophy and history—he could lose an entire day reading a biography of a long-dead president, whereas I might last two pages before I fell asleep. Philosophy was too abstract, too aimless for me. Mike would say the same was true of my work. For him, particle physics was witchcraft; advanced mathematics was sorcery. He would edit my writing for grammar and structure without being able to understand the meaning of a single sentence.

And yet: We were both capable of intense focus. We were both parsers. Neither of us accepted easy answers; we just asked different questions. It was as though our brains were the same machine, only wired for different purposes.

I would be leaving Harvard soon, and we had to decide what would come next. Our choice, it seemed to me, was binary: Either we were going to get married or break up. I gave Mike six months to decide between our possible paths. When Mike's time was up and he still didn't seem certain, I reminded him how long it had taken us to find each other. The odds of his finding another us were infinite. It was a life with me or a life alone, I told him. He stood in the light of our carriage house and chose me. I jumped out of my shoes.

We were married in the fall of 1998 at the University of Toronto. I didn't want a traditional wedding—I never believed that love needs an audience or public confession to be real—but my family wanted one. Mike and I stood in front of a small audience in Hart House, a beautiful Gothic building on campus. I'll never forget how he looked at me that day: with love, with hope, with pride, the way only a groom sees only his bride. After the ceremony, we turned to walk down the aisle, his hand in mine, and my hand in his, and I was surprised to see everyone rise to their feet and burst into applause. Our families and friends could see that our love story had reached its right and happy end. Our Unitarian minister chose to compare us to rivers, not moons: two streams that had run parallel until they finally, inevitably, fell into each other. I still thought of us as twin satellites. Formally joined in the eyes of the law and our loved ones, we remained, to my mind and I think to his, two separate pieces.

But somehow the pieces fit.

Late in my time at Harvard, not long before my PhD defense, I was invited to speak at the Institute for Advanced Study in Princeton, New Jersey, most famous for serving as Albert Einstein's academic sanctuary during and after the war. I was awestruck when I arrived. The Institute seemed like a place where monks should gather, convening in circles on the grass, under the branches of enormous trees. I was put up in an elegant bedroom in the corner of a mansion. The walls were all painted white, and the silence was almost eerie. It was a place for thinking about the universe.

I was met by John Bahcall, my host and the "boss" of the astrophysics group at the Institute. He was a legendary figure but didn't carry himself that way. His glasses softened his already kind face. His gray hair clung to his head in tight curls. He reminded me of a rabbi, and in a way he was one. In the 1960s, when most Americans were looking at the moon, John was enraptured with solar physics, determined to answer fundamental questions about stars. He wanted to know *why* the sun shines. Later, he was one of two people crucial to the existence of the Hubble Space Telescope, and the standard model of the Milky Way is named the Bahcall-Soneira Galaxy Model, half after him. He asked me how my walk over from the mansion had been, fatherly from the start. "If you were my daughter," he said, "I'd want to be sure the walk wasn't too far."

My work in exoplanets, like the field itself, was still incipient. I had decided to speak instead about my contributions to the timing of the atomic events that followed the Big Bang. My talk was in the library, and I took in the smell of the floor-to-ceiling bookshelves as I sat and waited for the audience to arrive. Postdocs and the occasional luminary began filing in. I pulled up my long hair and clipped it back, one less distraction. John soon arrived, and then in walked Jim Peebles, a giant in the field of cosmology. Today Jim is the Albert Einstein Professor Emeritus of Science at Princeton and winner of the Nobel Prize in Physics. He also happened to have made some of the original calculations that I had corrected.

I began my talk, again using an overhead projector to illuminate my work. I hadn't been speaking long when someone asked, a little abruptly: "How many electron energy levels did you use in your model hydrogen atom?" The answer was on my next plastic sheet.

"Three hundred," I said.

Another hand went up. "Did you include helium in your calculation?"

Once again the answer was on my next sheet, and I flipped to it. "Yes. Both helium and ionized helium."

A few sheets later, another question: "What is the specific reason the recombination of electrons and protons proceeded more quickly than predicted?" I would learn that this was the Institute style: Nobody is allowed to finish a thought unchallenged. Yet again the answer was on my next sheet. The audience laughed. I laughed, too. I looked across at Jim Peebles and saw him nod. That nod felt like acceptance.

The next day, John gave me a ride to the train station. I had no sooner slipped into the car and shut the door when he turned to me. "Sara," he said, "I'd like to offer you a job here."

I looked out the window for less than a trillionth of a second. I turned back to him with the widest smile. "I'm so pleased to accept," I said.

For years afterward, John joked about how quickly everything had unfolded, poking fun at me especially for not talking to Mike first. I should have. But the Institute just felt like the right home for me. In a field as vast and daunting as ours, mentors are so important. The best ones show you not only where to look but also how to see. I felt certain that exploring the universe with John would be as close as I could come to standing at the shoulder of Galileo.

The line between lunacy and scientific fact, John would tell me, is forever shifting. Former impossibilities become accepted truths, which means that astrophysicists can be judged only from a distance. To this day I don't know how much faith John had in the future of exoplanet discovery. Our research interests didn't overlap much, and he never told me what he thought was within the realm of possibility. He had been surprised often enough by physics, and by people, to have been humbled.

Maybe, as a young scientist, you have an idea that can't be proven despite its having a solid foundation in physics. It might make intuitive sense, but some theories, especially the most revolutionary, resist experimental evidence. John told us not to fear. When better instruments almost inevitably come along, and some future scientist, following your earlier hunch, uses them to make an important breakthrough, then your work was still worth pursuing. What you did this week or this month or even this year wasn't important to him. What mattered was the sum value of your lifetime.

What could make us more ambitious in our thinking? What could make us more daring? More and more, my life felt like the product of the decisions that I made. I could feel an unfamiliar certainty rising in me, that I was where I was supposed to be, doing what I was supposed to do, with the people I was supposed to be doing it with. It had taken me a long time to find my way, but for the first time in my life, I wasn't lost and lonely. I wasn't an electron anymore. I was atomic.

CHAPTER 4

In Transit

Einstein's oasis at the Institute for Advanced Studies felt more like a launchpad to me, the seeds of ignition in every blade of grass. I sat under those enormous trees throughout the fall of 1999 and pondered the next step in my journey to the farthest reaches of the galaxy.

Not long after I arrived, NASA sponsored an effort to find the first Earth-like exoplanet. By then, about forty exoplanets had been discovered using radial velocity. Astronomers were still stuck sensing the presence of other worlds rather than seeing them, and those they found were all too big and hot to give life a chance. The agency set a higher bar: It wanted to find a rocky planet of reasonable size that orbits its sun-like star in what's known as the "Goldilocks zone," neither too hot nor too cold to sustain life.

NASA also wanted evidence of that life's existence.

It was a virtually impossible ask, but NASA had sponsored countless unfinished efforts to find extraterrestrial life over the years. Now they wanted to wipe the slate clean and start again. They called the program the Terrestrial Planet Finder.

Four teams were chosen to participate; one included people from Princeton, and I was invited to join. I was surprised to learn that the belief that we could find another Earth was flourishing, at least in some circles. The truth was, we'd be thrilled to find a planet that was Earth-*like*. But engineers yearn toward specificity; they want to know what exactly they are designing their machines to do. We would tell ours that we wanted to find another Earth: a perfect copy, an identical twin. We planned to find another us.

From those first nights I spent looking through my telescope with my father at my side, I had wondered what else—who else—might be out there. I had always had the gut sense that we weren't the only light in the sky that life called home. Why us? We couldn't be that special. Now, for the first time in my embryonic career, I was being asked to help turn that feeling into fact. We needed to *know* that there are other Earths, to be able to point at a celestial map and say: *That one, right there*. That was our goal.

The word "no" was banned from our gatherings. David Spergel was our team's local committee lead, and we met every week at Princeton's Peyton Hall. A practically visible current leapt like voltage from one dreamer to the next, each new idea lighting up the room a little more brightly. For a brief spell, we had the budgets and youth to imagine a seriously fantastical future.

Apart from the ever-challenging problem of distance, we had to reconcile the fragilities of light. At its essence, astrophysics is the study of light. We *know* that there are stars other than the sun because we can *see* them shining. But light doesn't just illuminate. Light pollutes. Light blinds. Little lights—exoplanets—have forever been washed out by the bigger lights of their stars, the way those stars are washed out by our sun. To find another Earth, we'd have to find the smallest lights in the universe.

We began with a simple question: If aliens were observing distant exoplanets with their own version of the Terrestrial Planet Finder program, what would Earth look like to them? We all realized that Earth's brightness wouldn't be constant. The continents reflect light. The oceans absorb it. So if you were an alien who happened to look at Earth when North America was facing you, our light would be brighter than it might be a little later, when the Pacific Ocean slipped into dimmer view. Maybe we could use that curve in the light to infer the presence of oceans on distant planets, and thus water, and so perhaps a planet that might prove suitable for life. Oceans might be the biggest windows in the universe.

Given the chasm in relative brightness between stars and planets—a sun is about ten billion times brighter than an Earth—our main challenge was figuring out a new way to make the stars go dim. Our committee's most tangible achievement, credited to David, was a new design for an instrument to do just that.

Coronagraph: I love that word. It's an umbrella term for any kind of light-snuffing device built within a telescope. The first coronagraph was invented in the 1920s by a pioneering French astronomer named Bernard Lyot. He was studying the sun, and he had used two small, circular light blockers inside his telescope to create an artificial eclipse. His system worked well enough for looking at the sun. But, like the concentric ripples made by a stone when it's dropped in water, light waves radiated around Lyot's pair of internal shields. Those thin halos would obscure the far more distant exoplanets that we wanted to see. David suspected, and others on our team helped prove, that a shape more like a cat's eye would push those same light ripples farther out of sight. The darkness our coronagraph left behind wasn't a perfect darkness, but it was very, very dark.

Building a new coronagraph to go inside a new telescope might take decades, however. The Terrestrial Planet Finder program remained a product of our imaginations, not our hands. I wanted more immediate returns. Inspired in part by my committee work, my mind began wandering to those forty or so exoplanets that we had already found. Even if they couldn't harbor life, maybe they could reveal something about how to find it. It's easier to come up with a new way of seeing when you already know what you're looking for.

Theoretically, there was a way to find and study exoplanets other than radial velocity. If, for the moment at least, astronomers couldn't fight the brightness of stars, maybe we could use their power to our advantage. Bodies in transit sometimes align—not always, but every now and then. If we were lucky, it was possible that a planet might pass between us and its star. It stood to reason that the effect would be something like a miniature eclipse. The moon looks giant when it blocks out the sun. The Transit Technique, as it would come to be called, applied the same principle to exoplanets: We would find them not by the light they emitted, but by the light they spoiled. Nothing stands out like a black spot.

The Transit Technique made perfect sense to me. Working in conjunction with radial velocity, it could tell us more about any particular exoplanet than radial velocity could on its own. You can learn a lot about an object from its shadow. A few pioneering, risk-seeking astronomers began monitoring the most favorable of the few dozen or so stars that we already believed hosted at least one exoplanet, waiting for a transit. I called my former adviser at the David Dunlap Observatory to see whether we could try, too. The telescope's camera at my old haunt wasn't sensitive enough for the job, but Hot Jupiters were capable of eclipsing about 1 percent of their star's light— more than enough for us to measure with better tools. We calculated that each suspected short-period planet had a one-in-ten chance of passing in front of its star. Those aren't the best odds, but they are far from infinite. I woke up every morning wondering whether someone had detected a transiting planet in the night. I was filled with the feeling that the world might change with every email, with every ring of the phone.

It happened so fast, I almost missed it. That November, I decided to take a weekend off. Work had been threatening to overwhelm me; I was too caught up in the frenzy of potential discovery. I took Kira for a long walk across the Institute campus. Most of the leaves had fallen, and the weather was just starting to turn. I tried to breathe a little more deeply than I had been. I felt the muscles in my shoulders start to release.

On Sunday night, I checked my email. There was one from Dave Charbonneau, a student I knew. He had been at the University of Toronto behind me and was now doing his graduate work at Harvard. From the opening sentence, he sounded crestfallen.

Back in September, he wrote, he had collected data while testing his thesis adviser's tiny telescope in Colorado. The eventual goal was to be able to survey a wide field of stars in one swoop, increasing the chances that his adviser's team might find a transiting planet. For some reason, he hadn't looked at the data culled from the test star until November. When he did, it was a revelation: He had detected the first transit. He had seen a transit of a known planet—HD

209458b, a Hot Jupiter. It was absolutely fantastic news. He had erased the last shred of doubt that exoplanets exist.

Right around the same time, however, Geoff Marcy and Greg Henry, two far more established astronomers, had seen the same black spot on the same star. (Marcy was something of a celebrity in our field. He and his research partner would eventually claim seventy of the first hundred or so exoplanet discoveries.) Marcy, Henry, and their team had observed the star later in the season, as it was setting, so they saw only a partial eclipse. Dave had seen a beautiful full transit. I don't know exactly what happened next, but Dave's adviser might have confessed something about his discovery to Marcy. Rather than waiting to do his own follow-up, Marcy seemed to use that conversation to confirm his team's find. Within a week, they had issued a press release: The first transiting planet had been found, and they had found it. Dave Charbonneau had come in second.

Pioneers can be ruthless, I wrote Dave: *You are a fantastic scientist and time will show that.* Then I told him what he already knew. He should still publish his findings, regardless of what his competitors did. That paper would matter more in the end than a press release. *Time will tell the truth,* I wrote.

•

The more immediate truth was that one race was over, and a new one had begun. That walk across campus with my dog was the last break I would have for weeks. I had been turning over an idea—a genuinely original one—and the successful use of the Transit Technique gave it a greater urgency. I was practically swollen with it. A lot of science, especially pioneering science, relies on intuition. You'd be surprised to learn how many of our most significant discoveries began as a hunch, as a feeling. I didn't have any evidence that my idea would work. But I was doubtless.

I had realized that the Transit Technique might help reveal something more than the black silhouette of a planet. Immediately around that tiny, partial eclipse, the same starlight that was being blocked by an exoplanet would pass through its atmosphere. The starlight would reach us, but not the way regular starlight reaches us. It would be filtered, like water running through a screen, or a flashlight's beam struggling through a fog.

If you look at a rainbow from a distance, its many colors form a perfect union. But if you look at a rainbow more closely, using an instrument called a spectrograph, you can see gaps in the light, minuscule breaks in each wavelength like missing teeth. Gases in the solar atmosphere and Earth's own thin envelope interrupt the transmission of sunlight, the way power lines cause static in a radio signal. Certain gases interfere in telltale ways. One gas might take a bite out of indigo, while another gas might have an appetite for yellow or blue.

Why couldn't we use a spectrograph to look at the starlight passing through a transiting exoplanet's atmosphere? That way we could determine what sorts of gases surround that exoplanet. We already knew that large amounts of certain gases are likely to exist only in the presence of life. We call them biosignature gases. Oxygen is one; methane is another. Maybe the way to find Bigfoot really was by seeing his breath. We could start with Hot Jupiters, the planets we already knew, and their more easily detectable atmospheres. Like a skunk's spray, their traces of sodium and potassium would stand out amid the company of less potent atoms.

I kept my idea to myself, because I knew it was great— I was the first to see the potential of the Transit Technique for studying atmospheres—and now I knew, too, that great ideas get stolen. Dimitar Sasselov, my former PhD supervisor, was the only person I told about my theory, and he offered to help me bring it closer to practice. When we had worked out the details, I published a paper extolling what Dimitar and I called "transit transmission spectra." Reading the gaps in rainbows.

My paper received considerable attention. NASA was accepting proposals to use the Hubble Space Telescope; within a few months of publication, one team cited my work and won the rights to study the light that passed through the atmosphere of a transiting Hot Jupiter. I was furious not to be included on that team, which chose an older male scientist over me.

Within two years, their work revealed the first exoplanet atmosphere. It didn't surround another Earth, but my premise had worked. We had seen our first alien sky. I had profoundly mixed emotions on the day of the announcement. My theoretical efforts had triumphed, but I still felt as though I hadn't been invited to the party I had helped to throw.

News of the discovery was embargoed from the public until one o'clock on a Tuesday afternoon. I told John Bahcall that I'd be ready to talk about it at 1:01 P.M. Every Tuesday there was a formal lunch called (imaginatively enough) Tuesday Lunch, at which John took his seat at the head of a horseshoe-shaped table filled with some of Princeton's finest minds. I was proud to join them. John had a habit of cutting off boring speeches by clanking his glass, terrifying some of my colleagues, but I wasn't scared. He was won over by careful, quantitative work, and I had done the work. I stood up and explained what others had done with what I had done, the chain of succession of my idea, from my lamp-lit desk to deepest space. The thought made me breathless. We had traveled light-years. It hadn't taken any time at all.

John smiled with pride, and I warmed in his admiration. But he was far from satisfied. "What's next?" he said.

I was coming to understand that I would ask myself that question every day for the rest of my life.

•

Life doesn't just need certain gases to survive. It also needs certain solids. It needs a rock to stand on. Perhaps, I thought, transiting exoplanets could prove valuable in yet another way. Radial velocity gave us the mass of an exoplanet. By watching that same exoplanet in transit, and measuring how much of its star it eclipsed, astronomers could now determine its size. High school physics teaches us that mass divided by volume yields density, and given an exoplanet's density, we would have a pretty good idea of its construction materials: Dense would probably mean rock. We were still far away from *seeing* small, rocky planets, but here was a way we might home in on them sooner than we might with something like the Terrestrial Planet Finder. If astronomers could find a planet made of rock that also betrayed atmospheric evidence of biosignature gases, we could be looking at another us.

I turned my full attention to the avenues of discovery opened by the Transit Technique. I didn't have much interest in confirming the existence of exoplanets that we had already discovered. I wanted to do better. I wanted to find new worlds passing in front of new suns.

Like every investigation into space, the Transit Technique had its complications. Bodies don't often align. We would need to see the smallest passages, a tiny speck on an endless horizon, and we would have to see those passages repeat, as regularly as an orbit. That would require our looking at the same star for long periods. If aliens were using the same technique to find us, they would have to wait at least a year to confirm that Earth is a planet. (Closer to two years if they had just missed us the first time around.) The bigger problem, as far as my hopes were concerned: Even if we could somehow train a camera on an exoplanet during its transit in front of its star, what would we see? We'd see what we see of the moon during an eclipse. We'd see its silhouette, but nothing of its surface. We'd see black. I wanted a familiar blue dot in the sky. I wanted alien oceans, evaporating into alien clouds.

•

Happily, walking in the footsteps of Einstein at the Institute made me feel capable of miracles. Not coincidentally, I was also in a place where I could finally make a friend. I still had a hard time navigating social situations; once, dressed nicely for a work event, I said to Mike, a little excitedly, "I can pass for normal!" He smiled. "Yeah, until you start talking." Despite my being among so many like-minded colleagues, I sometimes saw things too starkly. I might change my mind about something, but if I did, I would swing from one extreme to the other. I rarely saw gray.

Gabriela was far from gray. A blond, blue-eyed Mexican astronomer, she was working on her postdoctoral fellowship, shuttling between Princeton and astronomy institutes in Chile. Gabriela and I soon saw each other as allies. We were both young, both ambitious. We both wanted to join the ranks of discoverers. Gabriela could also apply to use Chile's enviable ground-based telescopes, which, for me, was like finding out your crush happens to come from money. At the time, the Institute's offices were under renovation, and Gabriela and I huddled in a cloud-white trailer in a corner of the grounds.

Spending time with her was a little like spending time with Mike: We were two of a distant pair.

Gabriela was incredible at math. It flowed through her more innately than music; it was something like spiritual. She had a preternatural ability to design simple yet expansive algorithms, and her access to telescopes meant that she could collect new data to feed them. I was good at interpreting those numbers with my efficient computer code. Gabriela was the library; I was the reader.

We set ourselves the immodest goal of finding the first previously undiscovered exoplanet using the Transit Technique. Together we shared that tingle of anticipation known only to explorers, flush with the almost childlike joy that fuels every adventure. Like other astronomers trying to stake the same claim, we thought it best to use a telescope with a wide-field camera, capable of monitoring tens of thousands of stars at a time. We wanted to buy multiple tickets for the same lottery. Gabriela and I also knew that we wouldn't be able to find an Earth-size planet just yet. A new Hot Jupiter, massive with a short orbit, would still be a titanic discovery in every sense.

Gabriela and I began hanging out beyond work, too. Mike liked her, and the three of us had frequent dinners together. She even looked after our menagerie of pets when we went on another long Arctic canoe trip. Gabriela and I became a team and, with time, a successful one. She flew to the telescope in Chile and diligently couriered tapes filled with astronomical data to me. I ran the data through my new lines of code, based on her stunning algorithms. We felt on the verge of greatness.

There was one moment when Gabriela and I thought we'd found a planet. We were strung through with adrenaline. Astronomers all over the world were trying to beat us, and we were trying to beat them, but making a claim that proved untrue would destroy our careers. We looked at the data and had to admit that something wasn't quite right. At the end of our three weeks of allotted observing time, Gabriela flew home from Chile. We sat down in her office at the Institute to solve our dilemma. It was long past dark. Harsh fluorescent light reflected off our cherry tabletop as we scribbled away, pounding through algebra. Papers covered with equations piled up around us. Some fell to the floor. In that chaos we confirmed that we were seeing bodies in transit, absolutely. But the shape wasn't right; the numbers didn't add up. We looked at each other and didn't know whether to laugh or cry.

We came to the realization at the same time: We hadn't found a planet. We'd found a strange tangle of stars—what astronomers today call a "blend." One star had passed in front of another, but a third, nearby star was contributing just enough light to the equation to lessen the starkness of the eclipse. It made a star look the size of a planet. Our only victory was that we had kept our confusion private.

Gabriela and I never did succeed. During our shared quest, she answered frequent questions from an older, famous astronomer. She thought he was just curious, interested in our progress and perhaps impressed with our strategy. Later that summer we found out he was competing with us. He didn't steal our secrets; Gabriela gave them away. I was frustrated with her and with the situation—the other professor would have probably succeeded even without his duplicity. *Time will tell the truth.* Gabriela was devastated. I thought she would pull out of her dive and we would redouble our efforts in the ways that only she and I could. But she never recovered her enthusiasm for our project, and it felt as though she was also losing her interest in me. I felt let down. Abandoned.

In the fall of 2002, the older professor's team put out a list

of transiting planet candidates, including a previously unknown planet, OGLE-TR-56 b, in the constellation Sagittar-
ius. Another team used radial velocity to confirm the finding,
winning the accolades that come with being first. I spent two
days crying. My father was visiting and took me to New York
City for an afternoon. I told him of my disappointment while
we walked the crowded streets; he bought me some camera
equipment, as if to tell me that there were millions more
things still to see.

Then I had a conference in Seattle to attend, and Mike and
I took the chance to go on a long hike on Vancouver Island.
He didn't understand why being first in such an esoteric way
mattered. His inability to understand my sadness was mad-
dening, but he also helped give me a sense of perspective. My
work would have its highs and lows, but Mike would always
be there. Nothing had changed in the real world. I watched a
bald eagle fly through a river canyon in front of me and
thought: *Everything is going to be okay*.

In the middle of that race to find the first transiting exoplanet, my near-disaster with Gabriela led to a different kind
of achievement. The equations that describe the dimming of
starlight held another secret: They could be used to calculate
the density of a planet-hosting star, which would help us eliminate false positives, such as the effects of our mischievous
trio. We stayed up late one night and figured it all out—how
our wrong could help others make sure they were right. It was
hard for me to sit with Gabriela to write up our work for publication, staring into the heart of our dissolving partnership,
but we could still salvage something of our time together. We
published the lessons of our near miss, and that paper became
one of my mostly widely cited. It was the consolation prize I
received in exchange for the end of our friendship.

CHAPTER 5

Arrivals and Departures

I wanted to have children. I was in my early thirties, and I had always imagined that I'd have children one day. It would be almost hypocritical to dedicate myself to finding other life on another Earth and not bring new life to this one. But now it became more than a desire or an exercise in completeness; it was a need. I could feel my body demanding to carry a baby.

Mike did not share the feeling. Pets were one thing, he said. Children were something else. (I congratulated him on his observational skills.) I didn't see a lot of room for compromise. We had talked about children before we were married, and he had agreed to the idea in principle. Practice, as always, was a different matter. It was time for him to make another decision about what he really wanted in life.

Late in 2002, after three years with John in Princeton, I was offered a position at the Carnegie Institution in Washington, D.C., as a senior researcher. The Institution was founded in 1902 by Andrew Carnegie, a safe haven where scientists are afforded the resources to follow their ambitions to their limits. A celebrated astronomer named Vera Rubin had retired, opening a place. I asked John whether I should fill it. His job was to find his postdocs permanent positions elsewhere. We were all meant to be just passing through Princeton. He knew that, the way parents know that their children will one day leave them. That doesn't make their leaving any easier. I don't think John wanted to lose me. He didn't want to tell me not to go, either.

"Well, Vera did all right," he said.

That was his way of giving me one last blessing. I was on my way.

Before I could start my job, I had to get a green card. I couldn't leave the United States while I was waiting for it, so that summer Mike and I couldn't take our usual canoe trip north. We went to paddle the Grand Canyon instead. It was some of the biggest water I'd seen. My canoe went into a trough and I couldn't see beyond the crests of the churn around me; I was overwhelmed by the pure power of the river and retreated to the relative safety of our group's raft. Lava Falls challenged even our guide, who flipped his kayak. Mike, however, took it on in an open boat. He looked so skilled, he paddled so flawlessly, that strangers shouted their admiration to him from the shore.

Soon after Mike and I moved to Washington, I was pregnant. I was elated. I began having recurring dreams of a daughter with red hair and blue eyes, Mike in miniature. I didn't have any conscious preference for a son or a daughter, but those dreams were as vivid as my dream of our survival during that long-ago night on the esker: I was sure that we were going to have a girl. Mike began subscribing to my certainty. He wanted to name her Kira. He must have really liked that book. Since we couldn't have a dog and a child with the same name, we renamed our loyal, brindled canoe companion "Tuktu," the Inuit word for caribou. Then we waited for the new, hopefully less fearsome-looking Kira to join our growing family.

Out popped Max.

I was drained after thirty hours of labor, but I can remember so clearly when Mike first held him. Max was perfect, 10-out-of-10 on his Apgar score. His thick, dark blond hair looked as though it had been trimmed and combed by a barber. He had sky-blue eyes like Mike—my dreams had been accurate there—and they regarded each other with identical stares. They looked so alike in that moment. Their faces shared an expression that wasn't joy or sorrow or fear; it didn't even look much like love. They were both totally surprised to see each other. They shared the sheerest awe.

Two years after Max came Alex. (His middle name is Orion, the constellation that dominated the western sky on the night of his birth.) I had to concede that both of my children were indisputably boys. I wanted to have more kids, maybe a red-haired girl named Kira next time, but Mike felt that he had met his end of our non-compromise. Two children were more than enough for us to handle. Rationally, I could see his point—there is something scary about being outnumbered by your offspring, and there aren't many canoes built for more than four—but I had been so happy when I was pregnant. I glow thinking about it even now: all of that hope, all of that possibility, coming to life inside me. I tried not to feel wounded when Mike wanted to race off for the fastest post-baby vasectomy in reproductive history. I was upset, but his surgery put an end to our debates.

NASA was more open to further additions. In 2003, the agency launched another space telescope, Spitzer. It was a minor miracle of engineering. Unlike most telescopes, including Hubble, Spitzer didn't capture visible light; it detected infrared. That was important, because stars and exoplanets both emit infrared light in the form of thermal radiation. (A little more than half of our sun's light is infrared. You can't see it, but you can feel it. Infrared is why places get warmer in the sun than in the shade.) We still had no way to perceive exoplanets in wavelengths our eyes can see, since their stars remained too bright. But exoplanets, especially the biggest and hottest of them, compete a little better in the infrared. The best of their light comes in the form of heat.

Another race was on. I had taken a few months off after the births of both Max and Alex, but each time I grew itchy to get back to work. Mike and I hired a nanny to help fill in the gaps; I was still exhausted. I was caught trying to be both at home and far, far away.

Despite my desire for near-term results, I also still worked under the auspices of the Terrestrial Planet Finder and its grand, distant goals. While the original four teams had disbanded, a new team had started up, and I was thrilled to continue thinking about a space telescope with a built-in coronagraph. At an astronomy conference, an engineer named Charley Noecker invited me to hear about a tangential idea: a mammoth shield that would fly in tandem with a space telescope, working the way we might hold out our hands to protect our eyes from the glare. The shield would block the star, and the telescope would be able to see the smaller lights around it. That shield, an "external occulter," was fraught with technological tall poles. But two members of the presenting team—Jon Arenberg and Ron Polidan, both engineers from Northrop Grumman—were so enthusiastic, so confident in the future success of the project, that I couldn't help sharing their optimism. I joined their early efforts to make the scientific case for it.

After leaving Princeton, I continued to work on exoplanet atmospheres as well, but in a scattered, undisciplined way. The launch of the Spitzer telescope made me focus. I helped write a careful, punchy proposal—atmospheres and temperatures are intimately connected—and my team was awarded telescope time. The lives of space hardware are shorter than even our collective attention spans, and Spitzer cost hundreds of millions to build. Every minute such an expensive machine is given over to you is a gift. I felt an incredible sense of validation when, after years of planning and anticipation, Spitzer took my prescribed aim.

In 2005, I coauthored a study that detected the existence of a previously discovered exoplanet through its infrared. Planets that transit in front of stars can also pass behind them, in a secondary eclipse. When that happens, the combined light of the star and the planet drops a tiny bit, because the planet is blocked from view. Given that drop in brightness, we could measure a planet's infrared heat. Our findings made a lot of news. For the first time, we could *see* an exoplanet, albeit through invisible light. It was our old friend HD 209458b, which Dave Charbonneau had first seen in full transit. Now it had been seen three ways, making it one of the most researched planets outside our solar system. Spitzer couldn't photograph HD 209458b but could undeniably detect its presence. We had it surrounded.

Because we happened to be looking at HD 209458b's heat, we could also estimate its atmospheric temperature. That was my job. I came up with a pretty big number: a balmy 1,600 degrees Fahrenheit. HD 209458b is a great ball of fire, and obviously a little too warm to sustain life. The cooler signature of another Earth would remain a far more challenging find. Our progress was still undeniable, and I looked forward to expanding my work on atmospheres and transits with the Terrestrial Planet Finder's beautiful star-blockers. With those pieces in place, I felt sure we were finally on our way to seeing the invisible.

Then NASA suddenly postponed the Terrestrial Planet Finder work, and later canceled it entirely. I had been a proud member of the first generation dedicated to finding another Earth: We were going to become interstellar Magellans. Now we weren't even going to bother to look. I was crushed.

I hadn't learned to accept that the development of a piece of space hardware, like so many journeys, is rarely linear. It is murky and ugly and sometimes confused, a long war in which the advances hopefully outnumber the retreats until finally something gets won. National space agencies, NASA or Roscosmos or the European Space Agency, have their own priorities and agendas, set by governments that might rise and fall over a single budgetary dispute or illicit hotel-room rendezvous. One president wants to go to Mars; the next might settle for the moon. Giant corporations like SpaceX and Lockheed Martin have their own engineering departments fevering away. Universities develop and build their own technology for satellites. Every day across the world, thousands of smart, devoted people work toward the same ends—separately, in pockets. Dozens or hundreds of languages have words for *telescope*. Not many people know more than a handful of them.

That was the first time something I had been working on fell into astronomy's developmental morass, its promise unfulfilled. The lesson I took from it? The universe might be infinite, but our appetites for exploring it are finite, and so are our resources. Time is the most precious resource of all.

•

The Massachusetts Institute of Technology planetary science faculty called me while I was on maternity leave with Alex. They were inviting me to interview for a professorship. I was excited to make the trip back to Cambridge, not far from my former haunts at Harvard. I also had serious reservations. Before I'd taken the job at Carnegie, I had interviewed at Caltech. And Berkeley. And Princeton. And the University of British Columbia. And, as it happened, MIT.

None of those meetings had gone well. The worst one had been at UBC. At the start of my visit, I had to listen to some older male professors salivating over a crop of undergrads they had seen at an event the day before. With the exception of my hosts, nobody in the faculty had any interest in exoplanets. When they did deign to ask me about my work, it was about my early research in recombination after the Big Bang—something I hadn't thought about much since Harvard. They grilled me with a surprising hostility. Feeling bullied and insecure, I gave unsatisfying answers to every question. My last interview was with a junior biophysicist. He was the first kind face I'd seen in hours. "I bet this is the end of a very long day," he said, and I nearly burst into tears in front of him.

The fear at every school, palpable in the room, was that researching exoplanets was an intellectual dead end. Even among some astrophysicists, there can be such a thing as too much stargazing. A few dismissed finding exoplanets as "stamp collecting," an endless, meaningless search for new lights just so that we might name them. I couldn't convince the cynics otherwise. Despite the growing number of known exoplanets—by then about 150—people told me that I would never be able to achieve what I said I would. We would never see enough planets in transit to reach meaningful conclusions about them. The challenges would always be too great. My breakthroughs were mirages; my discoveries were flukes.

It had been a debilitating process. I went to work at Carnegie, doing my best to steel myself against what felt like an army of skeptics. There was a gap between our desires and our abilities, but there always had been, and we had always found a way across them. Then we used Spitzer to see HD 209458b. Unbeknownst to me, MIT had continued to watch me and my work from afar. By the time I went back for my second interview in early 2006, John Bahcall's line between lunacy and scientific fact had moved just enough.

This time, MIT offered me a job. Somehow, I wasn't sure whether I should take it. I was honored, but I had never been a teacher. I had never been responsible for the futures of students, only my own. There are a limited number of hours in a week, and my university duties would mean less time for my research.

It would also mean less time for my family. We had recently lost one of our founding members: Tuktu had died earlier that year from a brain tumor, taking her last breaths on her favorite couch. Mike had lifted her body from it and carried her away to be cremated. I cried every day for weeks after we buried her ashes. She was such a faithful companion.

I spoke to my father more than usual. We had always kept in touch, talking often and seeing each other when we could. He would still occasionally ask whether I'd changed my mind about becoming a medical doctor. When all of our conversations shifted to Tuktu and her death, it became clear to both of us that I wasn't prepared for such emotional work. For months after she died, my father and I inevitably talked about mortality. "Death is a part of life," he said. "We start to die the day we are born."

Now, with MIT's offer in my hands, I called my father again, this time looking for a different kind of support. He told me to take the job. It didn't matter what else was going on, what else might or might not happen. "When the door of opportunity opens, you have to go through it," he said, his voice rising.

•

My father might have believed that his journey toward death started the day he was born, but it began in earnest with a series of stomachaches. Through all of the many times we talked about whether I should work at MIT, he was suffering from a sporadic pain in his gut. I was at a conference in Chicago—Pale Blue Dot II—standing outside the Adler Planetarium, drinking a coffee and taking in a little sun, when he called with an explanation. It was windy and crowded around me, but the truth eventually worked its way through and landed like the verdict of a hanging judge: He had pancreatic cancer. In the wicked spectrum of cancers, pancreatic cancer is one of the deadliest. For most people, sickness means doubt. There was no mystery here. Pancreatic cancer carries with it the worst kind of certainty.

I left the conference early and flew to Toronto. My father was living in a new condominium by then, with a woman named Isabella, his common-law wife. He had never been short of girlfriends, but Isabella had become a permanent fixture. He was almost as restless when it came to his homes, but his new place was in North York, just down the highway from the apartment where he had torn that orange polyester blanket in two.

He looked broken when I arrived. He had already spent the morning summoning friends and trying to give away his things. He wanted to see his prized Rolexes strapped around other wrists, but no one wanted to take the treasures of a dying man. I gave him a long hug, and then we went for a walk together in the cemetery across the street. We had gone there for walks and bicycle rides when I was a girl. As the sun set it filled with the most beautiful collection of shadows. Now we walked again between the tombstones, and for once it was my help that my father needed. He was picking out his gravesite.

Midway through his career, not long before he began offering hair transplants, he had spent a couple of years working in palliative care. He had felt it necessary to take his turn at it. I think it was strange for him, becoming the sort of doctor who faced problems he couldn't fix. He had always felt in full command of fate, and not just his own. It's a different mindset when you give up fighting death and seek only to ease its arrival. That sort of honesty has its comforts but also its price. My father knew too much about what was coming for him.

I burned through my frequent-flyer miles shuttling back and forth between Washington and Toronto for the next few weeks. I brought Max, not yet four, with me on one of my visits, trying to lighten the load on Mike back home. One afternoon, I decided to get some air and take Max for a walk. My father rushed out the door after us.

"This might be goodbye," he said.

He had woken up to my leaving noises, and I think he was still in a dreamy haze. I told him that I wasn't going back to Washington; I was just taking Max for a walk. He wasn't about to take any chances. We looked at each other long enough for Max to get tired of waiting and start playing on the floor.

"Sara," my father said finally. "You are the joy of my existence. You are the best thing that has ever happened to me, and you have exceeded all expectations."

I was astonished. He had talked about our connection since I was young, but his style was indirect. He talked about love scientifically, abstractly—as though, like reincarnation, it was a force to which we were subject, as though we were passive actors. Love was something that happened to us. Now he was saying something different. He was saying that love was active, a feeling that we could order and measure, and that he loved me with all of the strength he had.

To me, there was something almost catastrophic about his revelation. It was as though my life and his life had played out in an instant in my mind, and only then did I realize: Nobody had loved me like him. His love, a father's love, was without conditions, without reservation, without peer. And only now, at the moment when I understood how big and rare that kind of love must be, did I realize how close I was to losing it.

My father hung on, and I hung on with him. On another visit Isabella and I took him to see his doctor. He looked truly sick. Tumors were starting to grow all over his body. One was forming over his eye, and his weight loss was significant enough to change the shape of his face. The doctor looked at my father and said, "Dr. Seager, you're at rock bottom. People hit rock bottom, and then they start getting better. I am not bullshitting you." I couldn't decide whether the doctor was a liar or incompetent. Then I couldn't decide which was worse.

My father didn't believe his doctor, either. We went back to his condo. He wanted to die in his home, not in a hospital. He soon developed a blood clot in one of his legs, visible under his pale skin. Cancer is so awful, an endless loop of invasions.

It can feel like the same set of burglars keep coming back to steal what little they didn't take the last time. At midnight, Isabella and I stood beside my father's bed. We both stared at his leg.

"If this blood clot gets to my brain, it's the worst thing that could happen to me," he shouted. Then he fell quiet. "If this blood clot gets to my brain, it's the best thing that could happen to me."

My father survived the night, and the next day, and the day after that. I went to California for work a couple of weeks later; I landed back in Washington to a voicemail from my father's brother-in-law, my uncle. He said it was time. "You don't have to come," my uncle said. "You probably don't want to see him like this." I left Arrivals, went home to repack my bags and book a new flight to Toronto, and returned to Departures. It would be years before I could walk through that terminal without tears.

When I got to my father's condo, my uncle greeted me at the door. "Whatever you do," he said, "don't cry when you see him." I've never been given more impossible instructions. It was more than my not wanting to say goodbye; I didn't know how to say goodbye. I wanted to beg him to stay. The only words in my head were: *Don't leave me.*

I went into my father's room. The tumor over his eye had grown large enough to close it. He looked alien. There were lumps all over his body. I tried to hold it together, waiting for him to wake up, but I failed. I put my head on his chest. I could barely get out the words.

"Dad, what am I going to do without you?"

He was being robbed of almost every faculty, but he could still hear, and he could still speak. "Whatever you usually do, of course," he said. He even managed to laugh a little.

"Dad, I mean nobody will ever love me the way you have loved me."

I had said out loud what was, in some ways, a terrible thing to say. But in that moment, I believed it. I *knew* it.

Now he didn't say very much. Now he said everything. "It's always so."

A parade of visitors came through during each of my visits. Doctors and patients. Family and friends. Some big, brawny men with heads of very natural-looking hair. Many of them were overwhelmed with emotion. Some of them cried. Most of them said something like, "I will remember you." Nobody says that to you until the end. Dying is the opposite of birth. Every former first becomes a last. There are new firsts, but they're all lasts, too.

Still my father hung on. It was early December, hard and gray outside. He had made it ten weeks or so, near the standard for pancreatic cancer patients. After several more days with him, I finally decided that I had to go home to see Mike and the boys. I had been away so much. I flew back to Washington and fell into a deep sleep in the early evening. I woke up to Mike's grip on my arm and didn't have to ask. Isabella was on the phone. I flew back to Toronto one last time.

My father had told his best friend, "See you on the other side." I couldn't help hoping that he was right about life after death.

Today I am largely estranged from my mother and most of the rest of my family. The bad memories are too close to the surface, I think, and we've all made mistakes in our learning to live with them. My sister, Julia, is the only one I see. But maybe . . . Maybe my father was reborn somewhere. I wondered whether I might meet his next incarnation. I still do. But what I knew, beyond any doubt, was that I would never see my

version of him again. The man I had known as my father was
gone. He was there, all my life, and then he wasn't.

●

Two weeks later, Mike, the boys, our cats, and I moved north
in time for the start of the winter semester at MIT. I was excited to return to the frozen rivers and happier memories of
New England, to Beth and Will's Christmas tree farm and the
pleasures of our relative youth. This time around, we could do
a little better than the carriage house. We found a pretty yellow Victorian in Concord, twenty miles and a world away
from Boston.

Despite the most inauspicious of starts, Mike and I had
finished building our own solar system, with its own discrete
centers of gravity: two boys and three cats living in a pretty
yellow house. He would continue to work from home, surrounded by his marked-up books. He also took over nearly all
of the practical duties that go into running a family. I was
never good at those things, and practice hadn't made me any
better. I still struggled to pump gas into the car; basic household chores mystified me. Mike agreed, in action if not by
marital contract, to take care of everything ordinary so that I
could focus on the extraordinary. A new, unassailable order
settled into my life, a productive simplicity. My place and my
objectives were clear. I took the train each morning and
watched the trees turn into concrete; at night the concrete
turned back into trees. All I had left to do was find another
Earth.

Concord is the home of Walden Pond, where Henry David
Thoreau sat in his cabin and stared across the water, reflecting
on the values of a well-lived life. One day, Mike and I stood
together on the edge of that same body of water, the first hints

of ice forming around its edges, a dusting of snow in the otherwise empty trees. The death of my father and the start of my academic career had coincided so closely, I couldn't help feeling as though, from then on, there would be our lives before, and our lives after. I was looking over the edge of a precipice.

CHAPTER 6

The Law of Gravity

A scientist's greatest fear is missing something obvious. Not making a mistake, because in science, the greatest advances sometimes come from trial and error. The danger is in not recognizing the opportunity you have in front of you. In 2007, the National Research Council published a report titled *The Limits of Organic Life in Planetary Systems*. It contained a single line that has haunted me since the day I read it: "Nothing would be more tragic in the American exploration of space than to encounter alien life and fail to recognize it."

When we imagine life on other planets, we tend to imagine life as we know it. We see trees and birds and rain feeding into rivers. We imagine that we need to find a planet with oxygen in its atmosphere, because for us oxygen is the essence of life. But for half of Earth's existence, there was no detectable oxygen in its atmosphere. For a long time, no oxygen was produced, and then the world absorbed whatever oxygen that bacteria pumped out until, after many tens of millions of years, molecule by molecule, a little oxygen managed to accumulate. On our planet alone, life has taken, and continues to take, so many different forms. Germs are life. Birds are life.

Elephants are life. Imagine that there is a planet out there with dinosaurs still prowling across its surface, but we miss them because we're too busy looking for little green humanoids. Even thinking of dinosaurs is too Earth-centric an approach. Maybe there's a planet that's only ocean, but that ocean is filled with schools of alien fish. Maybe there's a supremely intelligent life-form that lives for millions of years. Maybe it's become some kind of post-biological intelligence with hard drives embedded for better memories, and organs that self-replicate.

That all might sound far-fetched, like science fiction, but space affords us almost endless possibilities. We can't be intent on finding some version of us, or only some version of us. We might miss something different. Because we still can't really see an Earth-like planet around a sun-like star, some in our community believe that we should try to find *any* planet that might have liquid water on its surface. Venus has no life on it because it's too hot and its oceans have evaporated; Mars has no life because it's too cold and its only water has frozen into ice. The relative heat that a planet receives from its star is an important dictator of its capacity to host life. So what if we expanded our search to include super-Earths—exoplanets with a mass larger than Earth's but substantially smaller than ice giants—that orbit close to red dwarf stars, or stars that are smaller and cooler than the sun? They would be easier to see in nearly every way: They're bigger than Earth and closer to their dimmer stars. Could somewhere like that sustain life?

During my first days at MIT, I walked around campus and felt like I was inside a beehive. If I found a quiet spot, stood still, and closed my eyes, I could almost feel its collective energy rising through me. In one building, someone might be learning how to splice human genes. In that one over there, a new kind of robot would be taking its first tentative steps. New computers were being built. New materials were being engineered. Behind every door was another playground. It was a factory that turned imaginations into bigger imaginations. If clouds didn't exist, someone at MIT would be inventing them.

It felt like an entire city had been custom-built for someone like me. It wasn't just the dreaming going on there that made me feel that way; it was the quality of the dreamers. I wouldn't say there is an MIT "type," exactly, because people there love so many different things. But I had spent my entire adult life among academics and scientists, at places dedicated to mining the most from their talents, and still I had never been surrounded by so many single-minded people, the best kind of obsessives. The objects of their affection couldn't have been more varied, but the intensity of their affection burned at the same impossibly high temperatures. There would be no way to measure such a thing—if there were, someone at MIT would have figured out how—but I would guess that nowhere on Earth has a larger population of people who blink less. I've walked along the Charles River to get some air and seen dozens of people sitting on the benches that line the shore, together looking out at the water with the same unbroken gaze, seeing some crazy dream come to life in the wakes of sailboats. The rest of us will one day see their dreams come true. People at MIT build things, turning abstraction into practical magic. That's what struck me most when I arrived: This was a place where I could make something, something miraculous that I could hold in my hands.

I settled into my office in the Green Building. At twenty-one stories, it's a landmark in Cambridge. My office is on the seventeenth floor. Today my desk is piled high with journals and stacks of research. There are shelves lined with books: *Optics* and *Asteroids III* and *How to Build a Habitable Planet*. A blackboard hangs on one of my walls, and it's usually covered in scribbles. There are times when my office looks like a scene from *A Beautiful Mind*. Although I prefer chalk to grease pens, there are days when I blur the very fine line between advanced mathematics and abstract art.

On that first morning, my office was still empty except for the light that poured through its long wall of windows. I took in my sweeping view of downtown Boston and the ever-changing angles of its shadows, and I remembered one of the last conversations I had with my father, when I told him that I'd finally decided to accept the job at MIT. "At my age, it's the best I can do," I had said. It was probably the best I could ever do, full stop. I would be starting at MIT with tenure at thirty-six and deep into a stable academic career, and life, at forty. Progress in increments.

He glared at me. He might have been happy that I'd be taking the job, but he didn't like how I was approaching it. "I never want to hear you say that something is the best you can do," he said, surprising me with his passion. "I never want you to be limited by your own expectations." It had been his last lecture.

Sitting in my new office, I decided to think big, to try to make my brain like the universe, forever expanding. The whole point of tenure, after all, was to feel safe to pursue risky long shots. I affirmed the sense of purpose, the mission, that I'd spent my whole life searching for: I wanted to find another Earth, and then I wanted to find signs of life on it. That search had been considered most seriously under the Terrestrial Planet Finder mission, which was now on tragic, indefinite hold. Without it, I needed another way to see.

Years earlier, an astronomer had suggested that we might be able to use a new piece of hardware, the in-development James Webb Space Telescope, to read the atmosphere of a transiting exoplanet just a bit bigger than Earth, orbiting a star just a bit smaller than the sun. A planet like that would also be rocky. Something like us.

When I arrived in Cambridge, another fantastic new piece of space hardware was being developed: a telescope named Kepler—a new Hubble, a new Spitzer. It was the singular creation of Bill Borucki, a physicist who had embarked on an almost spiritual quest to build a space telescope designed to find transiting exoplanets. It was named for Johannes Kepler, the German mathematician and astronomer who became enraptured with space when, as a young boy, he saw the Great Comet of 1577. Bill wanted to give us that child's fresh eyes, seeing for the first time new lights in the sky. I didn't help get Kepler built, but I made a successful proposal to NASA to receive early access to its data, and I began counting down the days until its launch in March 2009, a little more than two years away.

At MIT, something called the Transiting Exoplanet Survey Satellite, or TESS, was also in the dreaming phase. A third promising instrument. I was asked to take part in its development. Like Kepler, TESS would search for transiting exoplanets, but it would take aim at different targets. New ways to perform the Transit Technique were being developed all the time.

The massive Kepler would look for Earth-size planets in Earth-like orbits around sun-like stars, thousands of light-years away. Unfortunately, its discoveries would be too far away for meaningful follow-up, for us to know if those Earth-*size* planets were also Earth-*like*. The smaller TESS would search around closer stars instead, favoring red dwarf stars that are "only" tens or hundreds of light-years distant. Because those planets would be nearer to us than anything Kepler might find, and because red dwarf stars are smaller than the sun, perhaps we would be able to monitor them with our next generation of telescopes to study planet atmospheres for signs of life. In other words, Kepler would give us potentially thousands of small worlds, helping to determine how common our potential doppelgängers are. If TESS became a real mission, its exoplanet haul would reveal fewer Earth-size planets, dozens at most, but it would give us greater odds of recognizing a rocky cousin with the potential for water rippling across its surface—and whoever might call it home.

I closed my eyes and tried to imagine what it would be like to live on such a different world. For an exoplanet orbiting a red dwarf star to receive enough heat to sustain life, it would have to be so close that it wouldn't spin on its axis nearly as quickly as Earth. It would be "tidally locked" like the moon, with one face forced to look always at its star, bathed in perpetual light, and the other always cast in darkness. With a star so huge in the sky, maybe the best place to live wouldn't be where it's always daytime. The planet's proximity to its star would mean bombardment with intense ultraviolet radiation and frequent, powerful stellar flares. Maybe, for alien astronomers, the best place to live would be where it's always night. Or maybe it would be somewhere along the line between the light and the dark, where it's always sunrise or sunset.

None of that seemed hospitable to me, at least within our species' narrow definition of "home." I understood why our community was investing in finding habitable planets around red dwarfs, but I wondered whether my fellow astronomers were looking at those planets only because it was easier, not because they might actually find something—like a person searching for her lost keys under a streetlight because that's the only place she can see. Deep down, I didn't want an approximation of Earth. I wanted to find an equal, better than the best.

Even with James Webb and Kepler on the horizon and TESS in our dreams, I couldn't shake the feeling that we were going to miss *something*. With my father's voice echoing in my head, I found myself thinking constantly about signs of life. I always came back to atmospheres, to gases, to the air that we, and someone or something else, might breathe. I reminded myself that alien air wouldn't necessarily look or taste or smell anything like our own. To find it, maybe we'd have to develop new senses.

•

My work verged on all-consuming. I felt a constant internal pressure to perform—to be the first, to be the best. I received accolades for my efforts. I won the American Astronomical Society's Helen B. Warner Prize for accomplished young astronomers not long after I arrived at MIT. (I was the first woman to be awarded it in more than fifty years, even though it was named for one.)

Mike received far less recognition for his supporting role. Nobody gives you prizes for getting your kids to school on time, or for making sure there are always diapers on hand. Whenever I took a moment to look down instead of up, my divided focus found Max and Alex. The boys weren't old enough to take with us on adventures, and they were still too young to leave for long stretches with somebody else. At work, I was a star. At home, only a whisper below the idyllic surface of things, Mike and I had begun a slow drift apart, as though

Mars had released its hold on its two moons. We both knew it. That kind of divide can sneak up on you, but once it's there its existence is hard to deny.

Mike blamed me, mostly. He said that he was the same person he always had been. He still liked to put a canoe on the roof of his car and find a foamy spring river to run. I was the one who had changed. Mike just wanted his canoe partner back, the person who spent her winters with him planning our next adventure and her summers in the same boat. Now, more often than not, he did the dreaming and doing alone.

My immediate answer was to hire help. Not long after we had settled in Concord, I scoured the online classifieds and saw one that stopped me with its sweetness: *Hi, my name is Jessica, and I'm 17 years old. I am a high school student in Waltham. I love kids, and* . . . She was looking for work after school. The boys immediately fell in love with Jessica. She was great with them, a warm bundle of energy, and I thought Mike would appreciate the lightening of his load. I'm sure he did, but it wasn't just more time that Mike wanted. He wanted more time with me.

I couldn't see a remedy. I never felt like I had enough hours in the day. Money was tight, too. Years earlier, I had set my sights on a new white-water canoe—a beautiful Dagger Rival. We had more boats than just Mike's Old Town Tripper by then, but I had become infatuated with the Dagger Rival. I looked for a used model with no luck, so I splurged on a new one, painted teal, its hull clear of even the faintest scratch.

Mike told me that it was a dumb boat, too big and flat for us. It was built for novice paddlers, he said, not for expert white-water canoeists. But only one of us was an expert anymore. He paddled most weekends; I joined him less and less, until eventually it was rare that I did. I had grown out of wanting to be cold and wet all the time, and someone had to look after the boys. There were springs when Mike and I hardly saw each other. I worked long hours during the week, caught up in the end-of-term rush, and he was gone on the weekends, chasing the melt, often in my Dagger Rival. We both had our passions. Unfortunately, as with too many married couples, the person with whom we had pledged to share our lives was no longer principal among them. It could feel as though we were misapplying our love.

There is a phase of your life when you're building toward something, when your entire existence can feel like a construction project, a to-do list that you'll spend years crossing off. Young kids, ascendant career. The universe is so enormous. But one day we'd be finished with our respective searches and we could find each other again. "We'll have time someday," I told Mike. "We'll have money someday. We'll have time and money someday."

I meant it, too. I meant it like a promise. Mike and I settled into an uneasy peace, pretending that the promise was enough.

•

Kepler was ready for launch in March 2009. Mike and I went down to Cocoa Beach, Florida, just south of Cape Canaveral. We brought along Max and Alex. I had some meetings to attend, but we made a vacation around them. We stayed in a hotel on the ocean with other space families. While the boys played in the water, I looked up the coast and thought I could see Kepler's Delta II rocket on the pad, shining like some distant skyscraper. It was hard for me to believe it was real.

The launch was scheduled for late at night. I found a babysitter for the boys. She came to our hotel, and she watched closely as I brushed their teeth after they were tucked into bed, using just a little bit of toothpaste so they wouldn't have to spit it out. She thought that was some trick, and I smiled at her apparent awe. I was about to watch a rocket launch a telescope into deep space, with which we might find thousands of new planets, but I agreed—brushing teeth in bed was a kind of miracle, too.

Mike and I headed for the Kennedy Space Center, where we waited in the Rocket Garden with hundreds of others, mostly scientists and engineers, a few of them with proud and buzzing families. The skies were clear but it was a little windy, the gusts blowing in the wrong direction. Instead of being able to watch the launch from the bleachers set up three miles from the pad, we were bused to a different viewing area, five miles away. I spotted Bill Borucki with his children and grandchildren, so near the end of one journey and the beginning of another. The rocket was bathed in spotlights, and it was all any of us could see. Even from our distance, it looked made of polished stone. It looked to me like a sculpture, like a monument to everything good.

We heard NASA's mission control over loudspeakers, making their way through their final system checks. We heard the word "nominal" again and again. At NASA, that's the word you most want to hear. Nominal means *go*. The rocket was a go for launch.

Bill and his colleagues had put decades of energy and effort, not to mention hundreds of millions of dollars, into Kepler. Now they risked being consumed by nerves and that ill-timed wind. Not long before, a different NASA mission had suffered a catastrophic failure, and a new satellite had plummeted into the Indian Ocean. There isn't a lot of middle ground with rockets. Every launch is all-or-nothing: Rockets either go to space or explode trying.

This rocket's engines sparked to life. Light covers five miles a lot faster than sound, and for an anxious second or two, there was only a bright, silent burst. Then a low rumble made its way through the swamps and into the balls of our feet. Our ears filled with the noise that fire makes. The rocket lifted off the pad, slowly at first, picking up speed as it climbed higher and higher, out of gravity's clawing reach. The solid rocket boosters detached on schedule, and we saw them glow red hot as they started their long fall back to Earth.

The rocket carried Kepler deeper into the night. Within a minute, the last of its light was gone. Kepler was in space.

•

We astrophysicists sometimes have problems with our perception of scale. Knowing that there are hundreds of billions of galaxies, each of which might contain hundreds of billions of stars, can make our lives and those closest to us seem insignificant. Our work, paradoxically, can also bolster our sense of ourselves. Believing that you might find the answer to "Are we alone?" requires considerable ego. Astrophysicists are forever toggling between feelings of bigness and smallness, of hubris and humility, depending on whether we're looking out or within.

In December 2009, I was invited to give the John Bahcall Lecture at the Space Telescope Science Institute in Baltimore. John had died in 2005 of a blood disorder. I'd been traveling when I opened my email; my heart dropped as soon as I saw the subject line: *John Bahcall*. I hadn't even known he was sick. Just like that, I had lost the source of my most unwavering support. John had been more than a mentor for me. He was closer to my father in science. He was so much like my dad: kind yet demanding, encouraging yet critical. He, too, accepted who I was—in all my focused intensity, my lack of social graces—without reserve or explanation. It struck me as odd that a man who could think so expansively, whose brain could wrap itself around the architecture of entire galaxies and the mechanics of stars, could be snuffed out by a microscopic fault in his blood.

Although I was determined to give the talk, and a second, public lecture at the National Air and Space Museum, I warned my hosts that I wouldn't be able to make it if anyone in my family got sick. I had reason to be concerned. Outwardly, Mike was his usual robust self, the picture of health, but like my father, he had been battling a series of small, strange stomachaches. Mike's doctor told him to take Metamucil. Maybe he was constipated? I had serious doubts about both the prognosis and the prescription. So I was adamant in my caveats, even though honoring John meant so much to me.

Mike was thankfully feeling fine when it was time for me to leave. While I was waiting in the greenroom before my talk at the museum, I reconnected with my host, a scientist named Bob Williams, perhaps best known as the person responsible for the collection of images known as the Hubble Deep Field. I'd met him earlier in the day, and now we talked some more. I was almost surprised to feel what I thought was a budding friendship with him, a man so accomplished, so filled with resolve.

Back in the mid-'90s, Bob was the director of the Space Telescope Science Institute and had held certain rights to Hubble. Over loud objections from many of his fellow astrophysicists—every minute that Hubble was in space represented an enormous window of possibility—Bob wanted to spend ten of his priceless discretionary days pointing the new telescope toward a tiny patch of Ursa Major, about the size of a penny held at arm's length. The widespread belief was that the patch was dark, dead space, absent of celestial bodies, and it would be almost tragically wasteful to stare at the definition of nothing for more than a week. Even John Bahcall ranked among those strongly opposed to Bob's plan. But Bob held fast, and Hubble took hundreds of images of that patch over ten consecutive, controversial days. The Hubble Deep Field revealed three thousand previously unseen points of light. Not three thousand new stars. Three thousand new galaxies. Bob Williams almost single-handedly discovered millions of billions of possible worlds.

"Stay in touch, Sara," Bob told me after the lecture.

Almost as soon as I got back home, Mike told me that he wasn't feeling well again. This time, his stomachache was different. It swallowed his body, and he collapsed into bed for twenty-four hours. I had a moment when I wondered, given my fears about missing the lectures, whether I had foreseen this turn for the worse. Mostly I worried what might be the matter with Mike. He rarely got sick. It was the time of year for the flu, and kids are germ factories. But Max and Alex weren't sick, and neither was I. Only Mike was sick. It wasn't the flu.

He got better for a little while. A week later, on a Saturday, he was struck down again. The pain was so intense that it made his forehead bead with sweat; his stomach seized with convulsions until again they released their grip. But almost exactly a week later, on the following Saturday, he was back in bed. I called his doctor. He told me to take Mike to the emergency room. I protested: "What *is* this?"

The doctor dropped his voice an octave. "Sara, this is serious."

I roused Mike. He was so lethargic he was hard to move. I bundled him and the boys into the car. I got Mike into the local hospital emergency room in Concord and began thinking about what I could do with Max and Alex. I didn't have anyone obvious I could call for help. By then, our regular babysitter, sweet and exuberant Jessica, had left for college. I didn't have any other friends in town. My colleagues at MIT were people I could ask for help in getting a satellite into space, not to look after my children, and my father's death had driven me further apart from my family than ever, spiritually and geographically. Desperate, I phoned the mother of a boy who was one of Max and Alex's best friends. "Of course, bring them," she said. I drove the boys over and tore back to the hospital.

By the time I returned, Mike was transformed. He wasn't in pain anymore, and he had the hint of a smile on his face when I walked into the room. I wanted whatever drugs they had given him.

The doctors took an X-ray of his stomach, and it revealed a mass. There were a lot of questions about what kind of mass. In the cavities around it, there was only darkness. It looked as though Mike's body were filled with ink.

We would need to wait for clearer results. It's weird, waiting for word like that. I called another sitter, Diana, calm and no-nonsense. We had used her before, but she'd taken a full-time job with another family; when I called the day after Mike's mass was found, I was surprised to hear that she was available again. I asked her to join our family. With the boys looked after, I sat with Mike in his room for most of the next couple of days. The rest of the world disappeared.

Mike lay wrapped in crisp white sheets while I sat in a co-coon of beeping machines. I wasn't sure what I wanted the doctors to tell us. Part of me wanted them to find something specific, so that we could go about curing it. I couldn't abide this kind of mystery. I wanted to find the unknown and make it known—whether that patch of space really is dark, whether we're alone in the universe. But another part of me was scared by the prospect of finding an answer. I was scared what the answer might mean.

A couple of days later, Mike's gastrointestinal specialist came into the room. He told us that Mike had a nearly com-plete blockage of his small intestine. "It's a large mass," the doctor said. "It might be cancerous. It might not."

Mike was sitting up in his bed, quiet and calm. If he was scared, he didn't show it. My mind immediately jumped to the worst possibilities. Here was my husband, with whom I'd shared Wollaston Lake and everything that came after, so strong and resilient and optimistic, and now all I could imag-ine was this apparition in his intestines taking him from me forever. I couldn't lose him. Not now. Not yet. I had made him a promise; I wanted to keep it. I began crying so hard that I lost my breath.

"Stop crying," the doctor said. He was controlled, but the way he said it, he might have been scolding a child for throw-ing a fit in a store.

I didn't stop crying. I cried harder.

"Stop crying," the doctor said again. "It could be nothing. He might not even need chemo."

He left the room the way a storm blows offshore. Mike and I stared at each other in his wake. Mike's eyes stayed dry. Mine stayed wet. In that moment, we looked at the same thing and saw two completely different meanings. Mike saw only the possibility of getting better. I saw another man whom I loved and who loved me, suffering from another stomachache. I saw no way out.

•

Mike was discharged from the local hospital on a Friday. The surgery to remove his mass was considered elective, not emergency. He would go on a special diet to keep his guts calm—lots of white bread and no broccoli, the opposite of what healthy people eat—and we could take a little time to find the right surgeon in the city. But we didn't even get a chance to start the search before Mike was back in the emergency room at the hospital in Concord. He had agonizing back pain. He had warned the doctors that he had partially slipped a disc in the past, and that too much time in bed wouldn't be good for him. They had ignored him, and he'd spent five days immobile during his initial diagnosis. Now he could feel his nerves screaming like wires down his left leg until they tied together in his ankle, where the circuits went dead. His ankle stopped working. The emergency room doctors sent him home with heavy-duty painkillers. I wondered whether he had lived the last of his life without medicine.

On a recommendation, we booked a consultation with a surgeon at Massachusetts General Hospital. His demeanor shocked me. He was slick and well-groomed, and he emphasized how mindless and routine Mike's surgery was for him. "Everybody gets the same surgery," he said, banging his hands on his desk at the end of the sentence. He did more than a thousand operations each year. We had to hope that the popular treatment was the best treatment. The surgeon told us that we could sign the contracts when we were ready, and off he went, presumably to open someone else's husband with all the detachment of a butcher disassembling a pig.

Mike and I looked at each other. We both wanted cool professionals on our side in his fight. But we also wanted someone who still saw Mike and me as human beings.

I complained about the experience to one of my colleagues at MIT. Her sister was a surgeon at nearby Brigham and Women's Hospital, and with a few calls we got an appointment with her for the next day. Dr. Elizabeth Breen seemed the philosophical opposite of the doctor at Mass General. She was a leading colon and rectal surgeon, but she still had her humanity about her. She talked about how closely she studied each of her cases and how careful she was in the operating room. She made surgery sound more like an art than a trade. We chose her.

Mike went in for his pre-op appointment alone. It was meant to be short and routine. I was on the train back to Concord after a long day at work when he called to tell me that he wouldn't be coming home that night. I could hardly make out what he was saying over the sound of the train.

"What? What are you talking about?"

Mike had been booked in for back surgery the next morning. His ankle was now frozen solid, and Dr. Breen wanted it fixed before she did her own work a week or two later, as Mike would need to be able to walk to recover. Everything seemed to be rising to a head.

I told Mike that I'd be right there. I could ask Diana to look after Max and Alex. But he said not to bother. It was rush hour. "Just go home and be with the boys," he said. "Come in tomorrow."

After I recovered from my shock, I remembered that I was

supposed to fly to Toronto the next day to give a talk to a student group at my old university. Obviously I had to cancel. I hated to do it; I had spent my adult life meeting every commitment I'd made. I despised feeling unreliable. But sickness doesn't care about your calendar, however meticulously your former self might have kept it. To alleviate my guilt, I called a professor at Toronto whom I had known since graduate school. He was a practiced public speaker, and I asked whether he could fill in for me. He asked me whether I was going to be paid.

"No," I said. I told him that I had agreed to do the talk for free.

"No," he said back. "I talk for money."

The professor was already on campus. He could stand up in the middle of dinner and give a talk that would leave you soaring. An hour of his time would have saved me a lot of anxiety when I was already feeling anxious enough. I needed a simple favor from someone I considered an ally.

"Seriously?"

In the weeks and months that followed, I would be surprised by how much a few tender hearts would give to Mike and me. Beth and Will, our long-ago neighbors, would take us in like family at their Christmas tree farm whenever we needed their gentle wisdom and calming scenery. But far more often, it seemed, people would disappoint me with their callousness. The professor wouldn't change his mind, and I took his rejection personally. I called and canceled my talk, and felt as though I had let people down. I also felt let down. Just as any one person, or even our entire planet, can seem insignificant with a change in perspective, what I thought of as my community suddenly felt cold and distant.

Mike was at the hospital, and I was at home. We weren't

alone together anymore; we were alone apart, and we were only in the opening rounds of our fight. Neither of us knew it at the time, but Mike and I had started parallel cycles, each vicious: For every potential cure, there would be a new concern. For every bandage, a fresh wound. For every kindness, a cut.

CHAPTER 7

Problems of Statistics

One of the great hurdles in looking for exoplanets is the time it takes to find them. The nearest and brightest sun-like stars are scattered all over the sky, which means that no telescope can take in more than a few at a time. But it's prohibitively expensive, as well as nonsensical, to use something like Hubble or Spitzer to stare at a single star system waiting, hoping, to see the shadows of planets we're not sure exist. Properly mapping a star system might take years. Bob Williams had needed only ten days to uncover the Hubble Deep Field, and still so many in our community had at first risen up in protest.

I had been trying to make a long-term plan to find another Earth. I wanted to invest myself in something. With the Terrestrial Planet Finder mission shelved and the James Webb Space Telescope on its way, the answer couldn't be another giant, majestic machine. I learned about what the community had taken to calling CubeSats—tiny satellites designed to a standard form, which supposedly made them cheaper and easier to build and deliver into space.

What if I made a constellation of CubeSats, each assigned

to look at only one star? I dreamed of space telescopes the size of a loaf of bread—not one, but an army, fanning out into orbit like so many advance scouts. Each could settle in and monitor its assigned sun-like star for however long I needed it to; each could be dedicated to learning everything possible about one single light. Hubble, Spitzer, Kepler—they each saw hugely. Maybe now we needed dozens or hundreds of narrower gazes, using the Transit Technique as the principal method of discovery. Earth might be ten billion times less bright than the sun, but it's only ten thousand times smaller in area. CubeSats wouldn't see what larger space telescopes could see, but they would never need to blink.

I talked to David Miller, a colleague and engineering professor who was in charge of what would become one of my favorite classes: a design-and-build class for fourth-year undergraduates. It was revolutionary when it started, because it was so project-based; after a few introductory lectures, the students dived into the challenges of making an actual satellite. I asked David whether I could use his class to incubate my CubeSat idea.

He was enthusiastic from the start. Maybe the best thing about MIT is that no matter how crazy your idea, nobody says it's not going to work until it's proved unworkable. And squeezing a space telescope inside something as small as a CubeSat was a pretty crazy idea. The main challenge would be in making something small that was still stable enough to gather clear data—a tall order because smaller satellites, like smaller anything, get pushed around in space more easily than larger objects. To take precise brightness measurements of a star, we would need to be able to keep its centroid fixed to the same tiny fraction of a pixel, far finer than the width of a human hair. We would have to make something that was a hundred times better than anything that currently existed in the CubeSat's mass class. Imagine making a car engine that runs a hundred times better than today's best car engine.

"Let's do it," David said.

•

My life became a study in contrast, the light and the dark, the hopeful and the hopeless. I spent my days at MIT with my students, trying to see. I spent my evenings at home with Mike, pretending to be blind. His back surgery was considered successful, but he would never again have full use of his ankle. My mind sometimes wanders to those doctors at the Concord hospital, how they had ignored Mike and how he had accepted being ignored, with the result a ruined ankle. I understand that astrophysicists might think in the longest possible terms. I get that we're among the few who count years in the billions. Still, I can't help wondering why we so often choose to suffer lifelong consequences in exchange for some short-term efficiency. Why would we ever trade temporary discomfort for a permanent one? That is the most impenetrable calculus for me.

Mike came home to rest before his second operation in as many weeks. At last the time came for his surgery. We returned to the hospital. Mike changed into his emasculating gown, and we sat together behind curtains that we pretended gave us privacy. If there had been real walls around us, I would have been climbing them. On the outside, at least, Mike looked stoic.

Dr. Breen appeared in her scrubs. She looked prepared and confident. "It's complex," she said of Mike's case, and I had a hard time understanding everything she said after that. I remember thinking her language didn't match her brave face.

She seemed to be lowering our expectations, preparing us for the worst. She was going to take out the affected segment of Mike's small intestine and the nearest set of lymph nodes. If Mike had cancer, those lymph nodes would tell us whether it had already started to spread. They would prove either the gateway or the firebreak.

Mike was wheeled away. I found a chair in the enormous waiting room. At its center was a big, curved desk, staffed by humans who apparently had been trained to act as though they had never been members of our species. They were surrounded by concerned families. The difference in facial expressions between the people working and the people waiting couldn't have been more stark. Those behind the desk looked like they were wearing masks. Those on the outside looked stripped of everything but their emotion.

I opened my laptop and tried to lose myself in a cloud of CubeSats. We would first call ours ExoplanetSat, before we finally named it ASTERIA. CubeSats, as a species of satellite, are much cheaper than regular satellites, because they're smaller and easier to launch; they take up a lot less room in the hold of a rocket, and it costs $10,000 to send a pound of anything into space. Unfortunately, their cheap manufacture makes them prone to failure. Many of them never work. We use the same hopeless term for them that doctors use for patients they never got the chance to save: "DOA."

One of our first hurdles, then, was a problem of statistics. (Every problem is a problem of statistics.) We needed to know how many satellites we would need to give us a reasonable chance of finding another Earth-size planet. Thousands of bright, sun-like stars were worth monitoring, but we wouldn't be able to build and manage thousands of satellites. We also knew that, given the ephemeral nature of transits, the odds of an Earth-size planet transiting a sun-like star were only about 1 in 200. Some of our satellites would also no doubt fail or be lost. If we sent up only a few, we would either have to be very strategic or very lucky to find what we were looking for. There was some optimal number of satellites that, combined with a smart list of target stars, would keep our budget reasonable but still give us a good chance of success. It was, in hindsight, a strange time for me to be calculating odds.

A few hours later—it felt like days and nights had passed—I heard my name being called over a loudspeaker. Dr. Breen came to see me. She looked tired. The surgery had been even more complicated than she had feared; Mike's guts were a knotted-up mess. His stomachaches came when food got stuck behind the mass, backing up like a clogged drain. But Dr. Breen had performed a quick biopsy of the tumor in the room and seen no obvious signs of cancer. We wouldn't know for sure until the lab did more tests. So far, though, so good.

Mike's brother, Dan, came to visit, and we took turns spending time with Mike in the hospital over the next week. I wasn't there when Dr. Breen delivered the verdict. Mike called me, the way my father had called me that day in Chicago: He had Crohn's disease, a painful inflammation of the digestive system. Its plague of ulcers had led to the construction of the mass. That mass was cancer. That cancer had spread to the neighboring lymph nodes. His cancer wasn't terminal, but it was Stage 3, one level removed from hopeless.

I hurried to the hospital. When I exploded into Mike's room, he was almost glowing, flush with his usual optimism, as though he were waiting for his turn to run a set of rapids. He wasn't sweating a single bead of concern. The news could have been worse, he said. Stage 3 is better than Stage 4. At least he had been given a chance.

"I just have to get better," he said, the way he might name a chore on his regular to-do list.

I just have to rake the lawn.
I just have to do the dishes.
I just have to get better.

Sitting next to his bed, I looked at him and marveled. What were the odds?

•

By February, Mike was back home, recovering from his surgery so that he could start chemo. He slept a lot. Diana had become devoted to our family, a permanent presence after school. But I still needed help in the early morning and in keeping the house in some kind of order. I put an ad online and found Christine, a kind woman in her fifties who professed a love for cleaning that I couldn't fathom. I told her that I'd need her for six weeks or so, the time we thought it would take for Mike to find his feet.

Mike didn't always see Christine the way I did, as a necessary savior. He saw her as a replacement, and he was sometimes mean about it. Only he could keep the kitchen organized properly, shop for groceries properly, or shovel the snow off the back porch steps properly. I was still coming to realize everything that he did. I didn't know a job even existed until it wasn't done. And it wasn't just what he did that had been a mystery to me, but how he did it. He had a particular way of making our morning coffee. I didn't understand which part of the machine went where, or how much water to put into it. I had no idea how bigger instruments like furnaces worked, either.

I spied Molly, Mike's big fat cat, in the bathtub. She looked confused, as though she had no idea how she had ended up where she had ended up. I knew the feeling. I lifted her out of her self-made prison but kept an eye on her. She didn't seem herself. I took her to the vet. Her liver was failing. The vet prescribed huge yellow pills that I had to stick down her throat, and now Mike and Molly shared their sickness as well as their company. It became clear that Molly, at least, wasn't going to make it. Only a month after I'd found her in the bathtub she was thin, close to death.

I told the boys that Molly wasn't going to be with us for much longer. I cried while I told them. In the back of my mind, I thought that how we managed her departure might matter. I wondered whether, like my father had done for me, I could serve as my children's principal illustrator. I wanted to show them that there was a right way to say goodbye. I told them that we needed to be generous and patient with Molly, to give her everything that she might want, love most of all. We took a lot of pictures. I told the boys that memories are important. All things live until they are forgotten.

One early morning, Molly was so still and quiet that I knew it was nearly time. I asked the boys to come upstairs to see her. They were still in their pajamas when they came into the room, and we sat with her. She was in our hands when she took her last breath.

Mike was sleeping and I didn't want to wake him. The fact that Molly was gone was one of the first things our boys knew before he did. They couldn't remember Tuktu's death; they had been too young. They were still young, but now they seemed suddenly older.

Mike would normally take care of whatever came next. I thought we should bury Molly, but it was winter, and the ground was covered in snow and too frozen for me to break. I didn't know what to do, so I wrapped her up and put her in the basement freezer. Like a morgue. It wasn't the most tender of measures, but it bought me time to think. Then I took the boys to Dunkin' Donuts for breakfast—a treat, to distract them. And to distract me. I dropped them off at school afterward, and then I went back to the house and waited for Mike to wake up. A few hours later I called the school, worried that Max and Alex might have been upset.

"No, they're fine," their principal said. "They just told me they'd gone to Dunkin' Donuts for breakfast." It wasn't until they came back home that they remembered to be sad.

•

Mike was ready to start his chemo. We went to the local hospital to meet with Dr. D., his oncologist. From the instant Mike had been diagnosed, I had done what I always did when faced with an unknown: research. I had embarked on my own course of study in rare cancers of the small intestine. I read medical review papers, following their citations to other papers, and following those papers to other experts. I emailed the country's leading specialist at the MD Anderson Cancer Center at the University of Texas, using my MIT credentials to get his attention. Academics build vast, intricate networks of knowledge the way ants carve out their caverns. Mike's cancer was so rare that construction had just begun, but I nosed my way down every tunnel. I ran headlong into every dead end.

Dr. D. had not practiced the same rigor. He seemed so uninformed, so incurious, that he filled me with a combination of rage and terror. I wanted to punch a hole in the wall. I wanted to throw a chair out the window. He didn't know what he didn't know.

I knew. I kept my knowledge from Mike, but I *knew*. The sample of people who had suffered the same cancer as Mike was small, and small sample sizes can lead to error. But the findings so far were clear. The median survival time was eighteen months; the five-year survival rate was zero. I asked Dr. D. some pointed questions. He looked alarmed and asked if I was a biology researcher, as though he realized he'd been talking to the wrong person the wrong way. I sensed the oncologist's fear, his discomfort plain even to me. He knew that I knew more than he did. I didn't stop pressing him. I was a scientist pursuing the facts.

"We'll wait and see," Dr. D. said. "We'll do our best."

He continued to speak to us as though we needed protection from the truth. Sitting across from him, staring at his implacable face, I remembered my father's doctor, the one who had looked into his only still-open eye and told him that he was going to get better, no bullshit. When did doctors start acting like faith healers?

I shouldn't have been surprised that Mike really liked the oncologist. He wanted only positive feedback, from as many sources as possible. He started attending group therapy, something he wouldn't have dreamed of doing before he got sick. He didn't tell me what happened there, and I didn't ask. He was cultivating a distance between us. I was never sure who he was trying to protect.

I tried to find my own version of positivity that summer. Not by believing that Mike would overcome his cancer, which I knew was out of our control, but by finding out how to live every day to its fullest. I became obsessed with the proverbial "bucket list." I told Mike that he should make one, and I thought about my own. I treated its manufacture like an experiment. I asked everyone I knew: "You have one year left to live. What are the things you want to do but haven't?" Most people gave similar answers: *Spend more time with family. Pursue work that's more obsession than job.*

I knew I had mastered only one of those.

I watched Mike for lessons. Ravaged by cancer, with his insides turned out, he could still be so strong. Not everyone can complete aggressive chemotherapy; Mike rode his bike to his treatments. That summer, in the middle of his chemo, he wanted to go on a canoe trip. His doctor moved one of his appointments to give him a three-week break. Mike flew to Idaho with a tour group and ran big-water rivers for days. I picked him up from the airport after, and he was tanned and rugged-seeming, with a little stubble and a worn baseball cap. He looked good. We'd been together more than fifteen years by then, and we had lived so much in that time. But at the airport, he looked the way he had when we were young.

In August, I asked him to take me and the boys on a canoe trip all our own. In my heart, I thought that it might be our last chance. I wanted Max and Alex to know Mike the way I had fallen in love with him, sitting tall on the water. Mike refused. He told me that he was spent, and the boys wouldn't enjoy it. They had whined so much on our last trip, remember? I told him through tears how important it was to me. He shook his head.

We did something we never would have done otherwise. We rented a town house in a ski resort in New Hampshire. It was the end of summer, so it was quiet. We drove up Mount Washington, which was dispiriting. Some people hike up one side of that mountain, while others drive up the opposite side. We had always been drawn to adventure in wild places—that's the side where we belonged. Instead we were driving to a summit on pavement.

The wind whipped across the crowded parking lot at the top. I was miserable, until Alex made an announcement. He was holding on to a signpost so that he wouldn't be carried away by the wind, the red pajama bottoms he wore all the time threatening to turn into sails. "One day I'm going to hike up Mount Washington," he said. "I'm going to set a world record." I wanted to laugh aloud, but inside I felt an almost overwhelming gratitude toward him. He was still so hopeful. He didn't know enough not to be. One day we would hike up Mount Washington. One day we might set a world record.

Mike and I made a careful, searching return to each other that fall. We shared a dark sense of humor, blacker than black. Mike dug a hole for Molly as soon as the yard had thawed. Wanting to protect him from proximity to death, I retrieved her stiff corpse from my makeshift morgue before trying to bury her. Somehow there wasn't enough earth left to fill the hole, and I had to drop a giant rock into her grave to make up the difference. "Where did the dirt go?" Mike asked, laughing.

Now Minnie May was starting to fade. Mike dug another hole. She might hold on till winter, and he didn't want another cat to end up in the freezer. He wanted to be prepared. He was also careful to leave the pile of loose earth right next to the hole, ready to be shoveled back into place. He wondered aloud whether he would be around to fill it or whether I might have to find another giant rock to drop on another one of our cats. Then he smiled, and I smiled back, and then one of us started laughing, and then the other one did. It had been so long since we had laughed like that, I forgot that the punch-line was whether my sick husband or my sick cat would be first in the ground.

In October, Mike had some more scans done, to see whether the chemo had worked. It had not. There were new tumors outside his small intestine, lodged in his abdominal cavity. The cancer had spread. Cancer that spreads during ra-diation is a killing cancer. Mike was terminal.

It was a Friday. I was supposed to fly to Italy that Sunday for a conference. "I'll drop everything," I said. Mike said no, I should go. He wasn't going to die tomorrow. *We're talking years, not months,* Dr. D. had told him.

"Years, not months," Mike repeated to me.

Dr. D. was either lying or ignorant. I looked at Mike and did one of the hardest things I've ever done. I nodded.

.

I came back from Italy. The next day, a miracle: Dr. D. called to say that he'd been wrong. Mike's cancer wasn't terminal. He could be fixed. There were new tumors, but they weren't cancerous. They were unrelated to the mass in his intestines. Dr. D. was going to send out the scans for more analysis, but everything was going to be okay. Whatever plans we had made, we could think about keeping them again.

That sense of escape, the feeling of relief—I'm not sure we'd experienced anything like it since that morning on the esker. We felt changed the way the fire scare had changed us. We no longer talked about some distant, brighter future, how one day we would cross off another item on our bucket lists. We were done with waiting for better days. We talked in the present tense. We talked about how we were going to make every correction and shed every curse. We had nearly lost so much because of our inattention, and then cancer had threatened to strip away from us the fragments of our lives that remained. Now we had been given a second chance. Not just Mike—I had been given a second chance, too. We had been given a second chance. Mike and I made a new promise, this one to each other: We were going to make everything right. We were going to take fresh stock of what was important and what didn't matter, and we were going to do that good work together.

After ten blissful days, Dr. D. called Mike back into his office.

He had some bad news.

He had been right the first time.

•

By November, Mike was preparing for more chemo. He maintained his optimism, seeming genuine in his belief that if his cancer proved fatal, it would not take him for some time. I kept my facts to myself. The night before the first appointment of his second round, he wondered whether he even needed to go. Not because he was going to die anyway. Because his cancer wasn't that bad.

I found his denial as frustrating as it was heartbreaking. Here was the evil of deception, self-deception especially. "The truth hurts," we'll say in one breath, and we'll call lies "harmless" in the next. That's the definition of bullshit. Mike's hope had become sinister to me. He needed to be thinking about how to spend the time he had left, not counting on time he would never have. I knew what his doctor was going to say when we saw him the next day. Even he wasn't going to be able to keep lying anymore.

"Mike," I said. "Do you want to hear the truth from me right now, or from your doctor tomorrow?"

"From you," he said. "I guess."

"You are terminally ill. We don't know how long you're going to live, but you're not going to live much longer."

He let out the heaviest sigh. Then he went upstairs and opened the children's bedroom door. He watched Max and Alex sleeping. In the dark and quiet, his eyes began to fill. "My biggest regret will be not seeing them grow up," he said.

Your mind plays a curious trick on you when tragedy strikes you or someone you love. You spend a lot of time remembering *before*. Your old life plays like a movie in your head. You remember when your worries were of a different scale, and you relive slideshows of happy scenes, every frame made gauzy and melancholic by everything that came *after*. Before catastrophe, you look ahead to give meaning to your present. After catastrophe, you look back. Mike no longer looked forward. He remembered. He remembered staring into Max's blue eyes for the first time. He remembered canoe trips and rocket launches. Now he knew what I knew. He looked at his sleeping boys and finally cried.

Mike still did the chemo. He was exhausted by it. One evening, he came downstairs to join Max and Alex and me for dinner. Mike pushed his plate aside and put his head on the table. He fell asleep for ten minutes or so, then woke up confused. "My head hurts," he said. He went back upstairs to bed. He hadn't touched his food.

I looked at the boys. I wanted to try to see their dad the way they saw him.

"So," I said. "Is your dad sick, or is he normal?"

They looked at each other, trying to confirm that they had the right answer before they gave it. They didn't say anything to each other. They only looked.

"Normal," they both said at the same time. Sick Mike had become the only dad they could remember.

•

Shortly after Christmas, we heard from Mike's doctor. The second round of chemo hadn't helped much. The new tumors weren't growing, but they weren't getting any smaller, either. Mike's cancer wouldn't be gone until he was gone with it.

On New Year's Eve, I put the boys to bed and sat with Mike at the kitchen table. We were different kinds of tired, and we had come to spend a lot of time like that: not speaking, just looking. I suppose we always had. The lights were mostly off, and the rest of the house was silent. There had been a lot of snow that winter, falling in reflective blankets of white. In the middle of the night, there seemed more light in our yard than there was in our kitchen.

"Wow," I said. "This was the worst year ever."

Mike looked at me the way he'd looked during the fire on the esker, forgetting to conceal his fears.

"Next year will be worse," he said.

CHAPTER 8

The Death of a Star

Everybody dies instantly. It's the dying that happens either quickly or over a long period of time. Mike spent a long time dying: eighteen months separated his diagnosis and his death. He was the perfect statistical norm; he followed the predicted schedule almost to the day. That meant he also conformed to every trope and stereotype that people use to describe a cancer sufferer's losing fight. In the ledgers of causes of death, Mike's would be filed under "long battle," not "instantly." For some reason we use "instantly" only when we're talking about someone who's been killed, usually in an accident. Sometimes the idea of "instantly" is used to underscore the tragedy of a situation: *He didn't know that today would be the last day of his life. He was taken so soon. We didn't get a chance to say goodbye.* Other times it's used as an expression of comfort: *At least he didn't suffer. He died doing what he loved. He didn't even know what hit him.*

I understand, intellectually, the need for the distinction. A car accident and cancer are two different strains of death. It's the difference between dying as a whole, all at once, and dying piece by lost piece. It's the difference between a building that's demolished and a building that's left to ruin. Either way, the building ends up gone, but the way it vanishes isn't the same, and we need a word to make clear the difference in process.

It still felt to me as though Mike died instantly. Yes, we knew his death was coming. We could get his "affairs in order," whatever hollow comfort that is supposed to bring, as though the most important thing when you die is that you die with a tidy desk. At least we could be considerate of the lawyers and accountants he would be leaving behind.

Though I recognize that his strength was a gift, there was also a special agony in how long it took for Mike to die, especially for him. It's never been one or the other for me, always both: We were unlucky to lose him; we were lucky to have him for as long as we did. I had one big regret, the promise I wasn't able to keep—the promise of our spending more time together—but at least I didn't have any of the small regrets that might haunt the family of someone who died without warning. We didn't get to do everything we wanted to do, but we got to say everything we wanted to say. I hate that. I am grateful for that.

The dying time that Mike and I shared didn't make his death any less of a horror, and it didn't make my loss feel any less sudden. Mike took a breath, and then he didn't. He was alive, and then he wasn't. In one moment I was a wife. In the next I was a widow.

•

One day in January, Mike approached me. "Sara," he said, waiting to make sure that I had looked up at him. "The doctor said I shouldn't die at home, because we have small children here." We had vowed to be honest in our conversations about death, and we had talked about where Mike should be when he died. Those weren't easy conversations, and I usually ended up in tears, but I had learned a lot from the way my father had wanted to die. I had been given a trial run at holding vigils; now I saw death the way I saw birth. Nobody's ashamed of talking about how they want to bring their children into the world, and I don't think anybody should be ashamed to talk about how they want to leave it. Mike and I had long agreed that, like my father, he should die in his own home—in our pretty yellow Victorian, with the boys and me. Now his doctor was trying to sow seeds of doubt in him. It felt as though we were about to start our death conversations all over.

"What?" I said. I could feel a white-hot anger coming to the surface. "What kind of lesson would that teach our children? That we dump sick people at the hospital to die? That's ridiculous. The doctor should know better."

Mike was silent. I knew he agreed with me. I started to cry; it was all so awful. I just wanted Mike to be home with us. I wanted to try to make him dinner and for him to sit up with the boys and for the last days of his life to be free from the buzzing of fluorescent lights and the hallway chatter of nurses. "We are going to teach Max and Alex that we will love you and take care of you until the day you die," I said. That was the last we spoke of where. All we had left to know was when.

Mike started a third round of chemo. I told him that he shouldn't, but he wanted to take any chance to survive. No part of him wanted to die. The chemo had no history of success, and the third round would be an attacking, experimental treatment that would kill Mike to give him a few more weeks of life. How dare Dr. D. do this to Mike? I protested, desperate to stop it. Even if it worked, what sort of life would Mike be leading? If he was going to die, I wanted him to have one last chance to feel strong, not defeated. I wanted us to spend time outside, the way we always had, or at least the way we had at the start of our lives together. The snow kept falling that winter, in feet rather than inches, and I wanted to go cross-country skiing with my bear of a husband, the sound of our poles finding the same familiar rhythms that our paddles once did. I wanted our last hours together to be like our first. The chemo would make that impossible. But Mike wanted to try. He always wanted to try.

That third round of chemo nearly finished him. He had to stop the treatment. Dr. D. had compared Mike's strength and determination to that of Marines and war veterans, but the pain was too much even for him, and it drained most of what was left of his spirit. I was so angry. There were nights when I thought about Mike's suffering and feared I might go blind with rage.

The snow continued to fall, January into February, February into March. If Mike wanted to do something remotely active, he knew that he'd have to spend the next however many hours laid out. He budgeted his energy, deciding whether something was "worth it." He had always wanted to go to the Galapagos Islands, the last in-reach item on his own bucket list. He saved up his reserves, and then he took a two-week trip with the help of his best friend, Pete. Not long after Mike came home, we held Alex's sixth birthday party at a gymnastics center. I have a picture of them playing together, tossing foam blocks at each other for five or ten minutes. Mike had to spend the next twenty-four hours in bed. It was worth it.

It seemed to rain every day when that spring finally arrived. What had turned into a record snowfall had melted, but the ground was still frozen, and the melt and the rain had nowhere to go. Concord locals called it a 100-year flood, the rivers overflowing their banks, closing roads and turning the low-lying parts of town into ponds. It was hard not to see the rising waters as a metaphor: We were powerless to stop what was coming. Mike had been relieved of the last of his denial. In the first week of May, he took one last trip with his canoe club to New Hampshire, but he came home spooked. He shouldn't have been driving. He had a hard time concentrating. He had always been so focused, but now a blurring fog had begun its descent.

There were still flashes of our dark humor. We continued to wonder whether he'd bury Minnie May in the hole that he'd dug in the yard or whether Minnie May would bury him. They were locked in a morbid race. I made a joke that at least I'd be able to have more kids after he died. "Sara," he said, "you need a baby like you need a hole in your head." I burst out laughing—my desire to have more children never really faded, but the chaos of our reality made Mike's assessment hard to deny. In a moment of sincerity, I told Mike that I'd wear only black after he died, the way Queen Victoria did after she lost Albert. Mike pointed out that it wasn't much of a pledge. I only wore black anyway.

•

I made up a new law: "The Law of the Conservation of Happiness." Conservation laws are fundamental to physics. The conservation of mass. The conservation of energy. The conservation of angular momentum. It's rare that something truly disappears. It might go missing, but it's still out there somewhere, in some new and hidden form. Whenever anyone asked me about Mike, I would turn away from the trauma. "You really need to savor every minute," I would say. "Life is short." Sometimes I said these things to a mirror.

I was still working hard, mostly on ASTERIA and biosignature gases. I was constantly torn in two, always some form of distracted. I checked in with Mike a few times each day— something I'd rarely done before he got sick—and came home early as often as I could. I was lucky to have a great group of graduate students and postdocs, and I leaned on them and their less complicated lives. I set one to work on ASTERIA's optics, another on precision pointing, a third on communications. Others continued to work on exoplanet atmospheres. I told them I wouldn't be around as much as I might have been otherwise. I stood outside in the cold with an engineering student and told him that my husband was dying. He said what people usually said: "Let me know if there's anything I can do." I heard those words differently than I once had. I took them as a genuine offer.

I resolved to build new bridges. Christine was in the house in the mornings, and Diana came after school, and I tried to treat them more like company than help. Jessica still came to see Max and Alex on weekends, too, sometimes taking them to her house for a visit. They always came back excited about the adventures they'd had, visiting a home without shadows. I wondered whether she might take a trip with the boys and me. I felt strange asking—it was a big leap to go from hanging out at the house to boarding a plane together for a family vacation—but she said yes. And so, while Pete visited Mike, the rest of us—Jessica, Max, Alex, and I—flew to Bermuda for a few days in the sun and on the sand. It felt like the start of building something new, the first glimpse of what our lives might look like afterward.

I was going to turn forty that summer. Mindful of the Law of Conservation of Happiness, I decided to host a one-day symposium in May: "The Next Forty Years of Exoplanets." A lot of scientists have a similar gathering at the end of their career, but why wait? It might have seemed a little self-involved, but I needed reasons to hope. I needed to think about the future and feel something other than dread. For a mission statement, I wrote: *We want to show our children, grandchildren, nieces, and nephews a dark sky, point to a star visible with the naked eye, and tell them that star has a planet like Earth. We will make this possible in the next forty years.* I wasn't the sort of person to host a party, but making it about the future of space exploration rather than about me made it more palatable. Plus, I thought people would be more inclined to come.

I sent invitations to a select group of colleagues. One by one, they accepted. Each RSVP was a burst of good feeling. Natalie Batalha from Kepler. Matt Mountain, then the director of Hubble. Lisa Kaltenegger, now the director of the Carl Sagan Institute at Cornell. Dimitar Sasselov, my former adviser from Harvard. John Grunsfeld, a retired astronaut. Drake Deming from NASA.

Geoff Marcy—the codiscoverer of seventy of the first hundred exoplanets—was one of the star attractions. I had asked our speakers to be provocative, and he took my request seriously. (I was appalled when he was later accused of sexually harassing graduate students and resigned from his position at UC Berkeley.) At the symposium, he climbed onto the stage and railed at the lack of investment in our imagination. "I'm unhappy about the last ten years, and the next ten years," he said. Despite the early success of Kepler—by then it had found more than a thousand candidate planets, waiting to be confirmed—the cancellation of the Terrestrial Planet Finder mission, back when I was at Carnegie, had demoralized all of us hoping to find an interplanetary twin. He declaimed against the lack of cohesion within our community, the infighting among astrophysicists that had so often stalled our progress. It was hard to move forward when we couldn't agree *how* to move forward.

Members of the audience jumped to their feet and started raising their voices back at him. At first, I watched with alarm: The last thing I needed to hear was another unkind truth. But over the course of the day, as speaker after speaker ministered with such passion, yearning toward discovery the way climbers talk about conquering mountains, I experienced a buoyancy so long-forgotten that it felt new.

Afterward, we gathered together on the roof of MIT's Green Building and made a champagne toast to each other. It was a beautiful early-summer evening, the spectacular Boston skyline just flickering to life in the twilight. The sky and the Charles River were the same shade of blue. For the first moment in what seemed like forever, I felt as though there was plenty of time.

•

I'd almost missed my own party, because Mike had been in the emergency room not long before. He was in cancer's final stages. He was in crushing pain, and I didn't know how to help him. We learned that he was eligible for home hospice. We were assigned Jerry, an older male nurse and one of my life's true saints. A hospital-style bed was delivered, and together we made Mike comfortable in the guest room upstairs.

I told the boys the absolute truth about their dad. They needed to know, but they hadn't needed to know until they needed to know. Every unburdened day I'd given them felt like a tiny victory, but it was time. We were on a train, going to spend the day at the zoo in Rhode Island. We were seated by ourselves in a little booth, a table the only thing between us. My heart was pounding, but I tried to look calm. I measured my breathing. I had been practicing this speech for months.

"Max, Alex," I said. "I have to tell you something."

I took a pause. They were listening.

"Well, it's not good news. Someone in our family is really sick. The medicine isn't working."

Another pause. I was looking into two sets of the widest eyes.

"I'm sorry I have to tell you about this."

Now the longest pause.

"Your dad isn't going to make it. He is going to die."

Alex almost screamed. "WHAT?" he said. "I thought you were talking about Minnie May!"

I got up and gave each a big hug and held their hands.

"I already knew," Max said.

We sat quietly for a long time. Nobody cried, but it felt like we were riding in a funeral procession. The train car was almost empty. The air was summer heavy. We rocked and lurched with every bump in the track, on our way to the zoo, practicing being a family of three.

•

Over those last few weeks, Mike got more and more confused. Sometimes he made sense. Sometimes he didn't. He had run the snack committee at the boys' school, and now he tried to work on the spreadsheet for parental assignments. He'd forgotten that he'd already handed the job off to somebody else, and school was over. Another night he woke up and went downstairs. I followed him into the kitchen. He grabbed a bunch of knives by their blades, as if to reorganize a kitchen drawer, and then he tried to cut something with one of their handles. He poured beer into the coffee maker. I was in a full panic, pleading with him to stop, when he finally stood still and announced: "I have no idea how to get back upstairs."

Jerry had told me about something called "terminal delirium." The brain starts to follow the body's lead toward failure. Jerry had seen it countless times. He told me that Mike would soon complain about strange things, like his feet being cold. Mike would likely seem completely lost until the moment before death took him. Then he would go crystal clear.

I didn't want the boys to see Mike that way. "Do you remember what we did for Molly?" I asked them near the end. "Do you remember how we took such good care of her and then we said goodbye? It's time to say goodbye." Alex covered his ears with his hands and ran out of the room whenever I tried to talk to him about Mike, but Max was ready.

He and I crept together into Mike's room. Mike summoned the last of his strength and sat up in his bed. He gave Max the biggest hug. "How was school today?" he asked. He had forgotten again that it was summer. He might not even have known. Max looked at Mike, and Mike looked at Max. It was so primal, it reminded me of the day Max was born, when father and son had first stared at each other with the same blue eyes. Now they looked at each other again, both smiling. They gave each other another hug. No tears. I'm not sure either of them knew that it really was goodbye.

Mike and I said our own goodbyes so many times. One time I was heading downstairs when he grabbed my arm. "You're the best thing that ever happened to me," he said. The weight of his confession made my lungs empty. Another time, before he was confined to his bed, he asked me to sit down. He had something important to say. It took him a little while to form the words. I was silent, waiting for him to find his cour-

age. "Sara," he said finally. "Sara, I know you will get married
again." Mike seemed to want to say more, but he couldn't get
another sentence out. The way you feel a change in the wind,
I sensed that he wanted to give me permission to move on. I
was floored, not least because new love was the farthest possible thing from my mind. I told Mike that even if I'd somehow known the outcome of our lives together, "I would want
to live these years all over again." I would go canoeing with
him on the Humber River, and I would go back to his place to
warm up.

Jerry thought Mike had days left. Days turned into weeks,
and weeks turned into more than a month. One July afternoon, Jerry took me aside and spoke in whispers: "I've never
seen anyone resist the way Mike is resisting," he said. Mike
really wanted to keep living. His fitness before he got sick was
one reason; his mental toughness was another. But Jerry
thought Mike was refusing to die because he was worried I
wouldn't be able to manage. "You have to stop asking Mike
how to do things around the house," Jerry said. I had been
barraging him with endless questions about how to do his
chores. Now I did my best to stop, but then I realized that I
didn't know how to get his canoe rack off his car. I asked
Mike, knowing that it was a challenge even for him, and
maybe there was a special wrench somewhere . . .

"It's too complicated to explain," Mike said. He fell back
to sleep.

Like my father before him, Mike hung on. He was confined
to his bed and rarely awake, but his heart still beat. I watched
him for signs of pain and dropped liquid morphine inside his
cheek several times a day. I worried constantly about his suffering. We wouldn't let an animal linger that way.

It was almost my birthday. Mike and I didn't usually acknowledge each other's birthdays, but I saw an opportunity. I went upstairs. Mike was sleeping. I lay down in bed beside him, waiting for him to wake. Eventually he stirred.

"I'm turning forty soon," I said. "The big four-oh."

"You know I'm not very good at remembering those things," he said. He said it so sweetly.

I shook my head. I was trying my hardest not to cry. I wasn't going to be okay when he died, but I wanted him to believe that I would be.

"Mike, you're the best friend I've ever had. We have been so lucky to have such a great life together. But I'm going to be okay. The boys are going to be okay."

Mike stayed quiet.

"Mike," I said. "For my birthday, as your last present to me, I need you to let go."

I turned forty. Two days later, Mike died in his hospital bed in our home, with me by his side. There wasn't a single tube to remove. It was the first time I'd helped build something beautiful by putting down my tools, and I don't think I've ever been prouder. Mike left the world the way he had entered it. That was my gift to him.

His last breath was his own.

- - - - - - - - - - - - - - - - -

CHAPTER 9

A Widow for One Month

There have been lessons I have chosen not to teach. Not all knowledge is power; not all things are worth knowing. Max and Alex never saw Mike's body. They did not see him leave the house.

It was late when Mike died. Max and Alex hadn't seen him for a few weeks, but they knew he was still upstairs in his bed. I wanted to shelter them from Mike's actual death, because I didn't want them to remember their father as a corpse. Mike's mother and his brother, Dan, were in Concord for the end of his life, and now I asked Dan to take the boys to the park. They had never been to the park in the dark. They had the place to themselves, and I think they felt a little bigger somehow, like they had been let in on a secret.

I kept my own secrets. I called the funeral directors, who arrived in their hearse wearing suits and ties. I had called them too early, it turned out. Before they could take him away, we had to wait for a hospice nurse to arrive to pronounce Mike dead. After she did, the funeral directors put Mike in a special body bag that's designed to slide down stairs. They carried him out of the house, onto the front porch, down the front steps, and into the hearse. They moved slowly, carefully. They offered no apologies, but they started the car as though trying to will the engine to be quieter than it was.

I had already ordered the box that would contain Mike's ashes. Dave, one of the funeral directors, had brought out some samples—and could tell that I didn't like any of them. Some were too plain and simple, almost slapped together. Others were too fancy, elaborate assemblies of turned mahogany. Dave asked me to tell him more about Mike. "He has simple tastes," I said. "I have simple tastes, too, but not as simple." I didn't really know; I had never given a lot of thought to the box that would hold my husband. I knew there was one right box, the perfect box, but I couldn't find the words to describe it. Not too cheap, but not too dramatic. Elegant but not precious.

"Leave it to me," Dave said.

Now I sat in my empty house and wondered what Dave had built. What box was sitting over there at the funeral home, waiting for what would be left of Mike?

The boys were still playing at the park. Dan didn't have a cellphone, so there was no way for me to let him know that the job was done. He had to stay out until he decided enough time must have passed. When they all finally came back, the boys were like fireflies. I tucked them into bed and could see their light through the sheets. I decided to wait until the next morning to tell them that Mike was gone, giving them a good night's sleep before they received their life-changing news. For tonight, they could believe that Uncle Dan had decided it was a perfect time to go to the park, and that their father was still upstairs, in the deepest of sleeps, rather than bound for a box I couldn't picture.

•

The neighbors knew Mike had died because the street in front of our house grew quiet. Mike's mother and brother left, and Jerry didn't need to come by anymore. Neither did the nurse's aide I had hired, or the attendant who had bathed Mike. The medical equipment company picked up Mike's bed. I got rid of the last of his pills and the rest of the apparatus from our war against cancer. There was a feeling of emptiness in the house, like how I imagine a battlefield must feel after the shelling has stopped. Part of me couldn't believe the damage that had been done. I almost couldn't take account of the loss. Another part of me felt liberated by the end of the fight. It was over, and I was still here.

Jessica took the boys to her family's house. I wanted them to be around her parents and sisters—a full, loving home—for a couple of days. When they came back they were never more than an arm's length away. They had been confused by the language of death: "Why do people keep saying we lost him? He's not lost. He's dead. Why do people say they're sorry? It's not their fault." They needed answers to hard questions. They even followed me to the third-floor bathroom in unspoken need. "Um, guys," I said. "I kind of have to use the toilet." But I needed to be close to them, too. Their room had bunk beds and another twin bed. Max slept in the twin, Alex slept in the bottom bunk, and I took over the top. There were five empty bedrooms in the house, but the three of us slept in one room, as though we were camping indoors.

I hadn't always spent enough time with my boys. As much as I loved Max and Alex, the same pushes and pulls that could keep me away from Mike also kept me away from them. Losing Mike made me want to close whatever distance had grown between us. I liked that I didn't think very much about anything else when I was with them. Young children especially can seem selfish, or at least self-interested. Their concerns don't go much beyond the immediate. That almost forces you to think on their terms—they demand that you keep them in view. Max was thoughtful, serene, with an incisive sense of humor. Alex was more attention-grabbing, often on the verge of saying or doing something outlandish. In their own ways they had both always been presences in a room.

After the boys came home from camp one perfect summer's afternoon, I asked Max if he wanted to play tennis, our principal shared interest. Alex could spend time with Diana. We didn't have to be home anymore, and I hoped that Max wanted what I wanted: to get outside, to luxuriate in our freedom from the machinations of sickness. I was excited when he agreed to hit the ball around. We got our things and began walking to the courts. We chatted away—Max wasn't usually so chatty—talking about nothing in particular, but talking to each other, and that was good. Max went silent for a little bit. It was such a beautiful day. Not too hot, not too cold. The Goldilocks zone.

"You know, Mom," Max said. "It's better to have a dad who is dead than a dad who is sick."

Hearing your eight-year-old say something like that stops you. It's like hearing them swear for the first time: You're shocked into an awareness that they've been taking in more of the world than you knew. I wanted to ask Max why he felt that way. Before I did, I realized that I already knew the answer. I understood what he was trying to say. By a boy's measure, he had lost Mike a long time ago. He had also lost me, especially near the end. Now Mike was gone, but I was back. For Max, that seemed like a fair trade. One was better than none.

I felt almost guilty for sharing Max's feeling—for savoring the strange relief that came with the end of anxiety, of uncertainty. I have always been a clear and focused thinker, but my thinking was never as clear and focused as it was in the weeks and months after Mike died. I don't know why. Grief had a way of crystallizing things, of pushing out every stupid thought about everything that didn't matter. When I went into my office that summer, the rest of the world disappeared even more completely than it normally did. There was just me and my work and an unbroken string of revelations.

One of my students was studying mini-Neptunes, planets smaller than Neptune but larger than Earth. We don't have any counterpart in our solar system, but Kepler was finding that they were the most common kind of planet in the galaxy. That drove home our need to think beyond our own experience in our search for other life. Another student worked to determine not just how we might calculate the composition of giant exoplanet atmospheres, but also how to use data from Hubble or Spitzer to measure in what quantities those gases exist. How much sodium? How much water vapor? It wasn't easy, but together we found a way, using the atmospheres in our solar system as stepping-stones and forming the foundation of what would be called "exoplanet atmosphere retrieval." It was a strangely gorgeous time of thinking. Mike's death had somehow made even the stars more luminous.

And yet: *Spend more time with family*. That was the answer everyone had given when I'd asked what they would do with one more year to live. My work was my passion, but I had never found the necessary gaps between my responsibilities to balance the other side of the equation. I hadn't been able to keep my promise to Mike. I would keep it with my sons.

•

I wrote emails to friends and colleagues. The subject line was *The End of a Long Journey*. To some, I extended invitations to Mike's memorial service. To others, I passed along my eulogy or wrote more personal notes. To a rare few, I made a request. I wrote one of those special notes to Riccardo, an MIT alumnus and someone I had gravitated toward after the death of my father. I was drawn more and more to mentors who talked to me the way my father had, the way John Bahcall had, with no motive beyond their wanting the best for me, for my life to be good—which is to say, as though I were their daughter. *I am gradually turning my attention to new beginnings, and I am really hoping you will play some part in this.* The replies poured in, and I couldn't bring myself to read many of them, but I made sure to read Riccardo's. *I and all your friends continue to be alongside you,* he wrote, *for the rest of your life.*

I wanted to teach my kids that not only does life go on: It can still be an adventure. There were still so many unknown places to go, so many unseen things to see. I wanted us to go to Egypt and visit the Great Pyramids. Unfortunately, the new reality of my being a working single mother, along with the tumult of the Arab Spring, meant that Egypt wasn't in the cards. We could only manage a couple of days in New Hampshire.

We stayed at the Indian Head Resort, named for the face of the Native American that people claim to see in the contours of a nearby mountain. The boys loved it there. There were pools, an arcade, and endless hiking trails winding through the trees. I felt relief at this vacation from the sadness of home. But I looked worn, close to strung out, and other guests stared at me.

It didn't help when I had a meltdown in the resort restaurant. I'd gone to make a dinner reservation, and I approached a young woman who stood behind a book of what I assumed were names and times. "I'd like to make a reservation for six o'clock," I said. She shook her head and said they didn't take reservations; it was first come, first served. The boys had gone down to the arcade, and I was anxious to join them. "That makes no sense," I said. I imagined us showing up and having to wait, hungry and tired. "Why don't you take reservations?" My desire to return to a scheduled life ran hard against my current inability to function in polite society. I had lost whatever limited talents I had for pretending that I belonged, and I became unglued. That poor woman looked terrified of me. She called for the manager, who ended up looking nearly as scared.

It was still a getaway, and I tried not to feel guilty for reveling in our strange feeling of release. The boys seemed equally weightless. The hikes we took reminded Alex of his pledge to climb Mount Washington. Earlier that spring, we had done some hard hikes in the White Mountains, encouraged and accompanied by Brice, a Swiss postdoc on my team at MIT; he'd carried Alex over the most challenging downhill sections of the trail. Now, Alex and I began stealing days here and there to train, using nearby Mount Monadnock and the Wapack Range for practice runs. By August, nearly a year after we'd driven up the one side of Mount Washington with Mike, we were ready to climb the other side, just the two of us.

We made the three-hour drive the day before our planned hike and checked into a nearby hotel. The pool was filled with perfect families, splashing in the water: fathers, mothers, and their children, playing together with an abandon that my boys and I would never know again. I felt another meltdown coming, but I didn't want Alex to witness one up close—not there, not then. I fought to keep my anger from spilling over. *I hate you*, I remember thinking about those splashing families. *I hate you and your happiness*. They kept splashing away.

Luckily I had Alex and a summit to attempt. We woke up early the next day and wolfed down stacks of pancakes. Then we drove to the base of the mountain. The sky was cloudless. The sun was high. We bathed in infrared.

We decided to go up the shortest, steepest path, four miles long with a nearly 4,000-foot gain to the summit. It followed a creek. Alex was practically running from the start—he hadn't forgotten his pledge to set a world record, either. I almost couldn't keep up. We passed hiker after hiker, bolting our way up the slope. Alex was acting as though he was on the biggest adventure of his life, which he was. After a couple of hours, we reached a hut where climbers could stay overnight if they needed to break up their journey. We stayed for an hour or so, just to catch our breath and enjoy some hot chocolate. Alex was still only six years old, and his little legs must have been burning. But we kept going, hiking for another hour past the treeline, onto the rocky shoulders of the mountain. The view was magnificent. We could see a hundred miles, and soon the summit was within reach. Alex stopped not far from it and looked out at the country that we had won.

"Live your dreams, face your fears, and pay attention to your surroundings," he said.

To this day, I have no idea where he got that from. And he has no idea what he did for me. Mike's death was so many terrible things, but in the most unexpected way, watching him go had made me want to live the biggest, best life I could. I told Mike as much before he died: I told him his death would inspire me, that in his name and the pain of his loss I would never waste a single day. I would fight to do amazing things. I would be even more determined to find another Earth, my hope more like a mission, and I would help my boys find the same sense of purpose.

Now here were Alex and I, nearly at the top of a mountain. We practically ran the rest of the way. He didn't set a world record, but I was proud of him, and I was proud of myself. Maybe we weren't a perfect family anymore. We were still a family, and we could still make happy memories. We took a train down the mountain and drove back to Concord. That night I slept in the top bunk with the boys, more deeply than I had slept in years.

●

Jessica had transferred to a school closer to home, and she agreed to move in with Max and Alex and me. I needed her help with the house and the boys, but I also needed to fill the adult-size hole that Mike had left behind. She wouldn't pay rent; I wanted our home to feel like her home. I set about making her a suite on the second floor. She chose lavender for her bedroom color, and I hired contractors to convert the laundry room into a bathroom that would be just for her. A fresh start.

The contractors brought in a Dumpster, and it sat in the driveway during the demolition. For some reason seeing it sparked another meltdown. These were the strangest episodes, coming almost always without warning. Sometimes I was alone when I became overwhelmed with emotion; sometimes I had an audience of horrified faces around me. The only commonality was my powerlessness to stop them. My voice would rise into shouting, screams that came out of me like steam until my hair clung to my tear-damp face.

This time I was mad at Mike. Not long before he died, we had spoken about which of us had been given the worse end of a horrible deal. The one who gets sick and faces an early death? Or the one who endures the trauma of watching someone she loves suffer and die and then is left to pick up the pieces? I didn't know the answer, but suddenly it felt as though Mike had abandoned the boys and me, as though he had decided that death was preferable to living the rest of his life with us. I wasn't being rational or reasonable. Mike didn't choose to go. He had fought so hard to stay. But now he was gone, and my grief surfaced in another rage.

Mike had been something of a hoarder, and I could be sentimental about objects, too. We had fifteen different canoes and kayaks. We also had *pieces* of boats—boats that had been torn in half by rapids, fragments of shattered boats that hadn't even been ours. Mike had closets full of old clothes that he never wore, and the garage was filled with rusty tools. There were piles of old manuscripts, the blueprints for textbooks long obsolete. Now I looked at our bedroom furniture and couldn't stand it anymore. The cats had torn apart the fake-leather frame that held a battered futon. The bedside tables and dresser looked banged-up and worn.

I began throwing his things into the Dumpster in a frenzy. I didn't have Mike anymore; I certainly didn't need his stuff. A Dumpster can hold a lot. I filled a big chunk of it. The contractors worried that they wouldn't have enough room left for their needs, but they didn't dare stop me. I threw away our entire bedroom set. It felt as though I was shedding an unbearable weight, in what was part eviction, part exorcism. I had no desire to live in the past; no part of me felt in need of reminders. I had an insatiable appetite for space.

•

That November, I went to Newbury Street in Boston to get a haircut. It's funny, the things that come to feel like luxuries. I wanted to sit in a chair and zone out for a little while. A haircut was a good excuse to do it.

On Newbury, rows of old brownstones have been converted into stores, one on each floor. I've always found the entrances and stairways confusing. Instead of going into the hair salon, I walked up the wrong flight of stairs. I found myself in a messy room, filled with stacks of white paper and yellow file folders. A tall woman with blond hair and glasses approached me and asked if she could help.

"I was trying to go to the hair salon," I said.

She was about to point me in the right direction when I figured out that I was standing in a lawyer's office. I hadn't taken care of the legal affairs that follow the death of your husband. They seemed insurmountable and, honestly, not all that pressing. Death has a way of making everything seem urgent and unimportant at the same time.

"Do you know anything about wills and stuff?" I asked the woman.

Her name was Freya. She took me in with her eyes and nodded. Then she led me into an adjacent room, small and tidy and private. She told me that she handled all aspects of family law, because her clients, mostly businessmen, were always getting divorced. She somehow sensed that I wasn't getting divorced. I was staggered when she volunteered that she had been widowed ten years earlier. Freya had been through the process as both the lawyer and the client. Of course she would be happy to help me. By then I had almost stopped listening, stuck trying to figure out how she knew that I was a widow. I wondered whether she might teach me the secret handshake.

"How have you been doing?" she asked.

"Extremely well, actually," I said. I told her that life wasn't always easy, but I had been feeling empowered. Even my temper felt like a kind of justice. As much as I missed Mike, I didn't miss the last eighteen months of his life. "Liberated, almost," I said.

Freya smiled. She said that my euphoria wasn't unusual. I had earned my almost manic approach to life. It was a bubble of self-protection, she started to explain. A routine psychological response to trauma. *Oh no,* I thought. *I don't need to be hearing this.* I started backing away, literally, the way you retreat from a crazy person who wants to pray for you on the subway. Freya was not dissuaded. She started telling me her own story. Ten years ago, she had felt like a superwoman for months. Then her feelings of release were replaced by a withering aimlessness. She said that would happen to me, too. There would be a moment, as inevitable as death itself, when I would feel not just alone but lost. She wanted me to be ready for it. I had a better chance of surviving if I knew it was coming.

I didn't really know what to say. I know how I felt: I felt as though there are some things you don't tell people, partly out of common courtesy, and partly because they will never believe you. Widowhood, the birth of a baby, death itself—you have to let people experience those things for themselves, in their own particular ways. I understood better how Mike must have felt when I needed him to accept his own reality, when I needed him to reconcile his defiance with his odds. There was nothing about my experience that was boilerplate. Why would all widows be the same? Why would our passages be universal? We weren't the same before our husbands had died. Surely our stories were as diverse as we were. Maybe Freya had fallen into an impossible blackness—that's what she called it: "an impossible blackness"—but that didn't mean I would. Why couldn't I be fine? I wanted to be fine.

I stammered out a thank-you and said I would be in touch. She smiled at me, a little sadly, and shook my hand.

"I'm going to leave now," I said. I was going to get my haircut. I was going to sit in a chair and zone out for a little while.

I stepped outside and felt my legs go out from under me. *Whoosh*—it was as though I'd stepped back onto the street and been lifted into the arms of a storm. I somehow knew in the instant that I wasn't going to be one of those miracle babies, deposited by a tornado into the branches of a welcoming tree. I wasn't going to be set down gently. If the storm ever did put me down, it was first going to carry me far from the new world that I had been trying to build for myself, far from my illusion of happiness.

Or maybe it had already picked me up months before, in the moment when Mike had died. Maybe it had taken Freya to point out that I had never been defying gravity: I was the victim of it. Maybe she had known that I was a widow the second I had walked through her door because I had mistaken down for up, floors for ceilings. I had confused falling with flying. Surrounded by an impossible blackness, how could I have figured out the difference on my own?

CHAPTER 10

Impossible Blackness

Astronomy forces you to look at the universe with different eyes. We normally find things by looking for them. When we lose something, we keep our eyes peeled and retrace our steps and search until we find it. That doesn't always work in space. There is too much darkness, and there are so many places we have never been.

Sometimes we identify things through the absence of something else, the way the missing pieces of rainbows betray the existence of certain gases. Other times we find things through their effects on something else, like that telltale gravitational wobble that an orbiting exoplanet might cause in its star. Nothing else could be massive enough to make a star move, so there must be a planet somewhere nearby.

And sometimes we find things by studying that which can't exist on its own. The thing you're looking for must be there, too, because it had to have come first. If we find a table with four chairs around it, we can deduce that four people must sometimes sit around it, because why else would there be four chairs? Astronomy is haunted by the presences of things we can't see. Astronomy is like loss that way. It's like love.

•

My work sometimes suffered during Mike's sickness. Even at something like my birthday party I was of two minds at once. I missed a deadline to extend my Kepler access that I didn't even know was coming, and I lost my hard-earned rights to early data. I called my friend Riccardo. "Sara," he said. "You will stop crying. You will move on." Kepler was finding possible exoplanets at a rate of about one per day; in the time it took pioneers to conquer a few miles of the expanding West, astronomers saw whole new worlds. Now I would watch that progress unfold from the outside, feeling even more alone than I already was.

But after Mike's death, in that strange window of perfect clarity—in the right frame of mind, on the right rainy afternoon—I could escape again into the twin expanses of deep space and my dreams. My proposal to study exoplanet atmospheres using transit transmission spectra was now close to standard practice. Dozens of Hot Jupiters and their alien currents had been observed using Hubble and other telescopes. It was gratifying to lay claim to such a tangible contribution to the field. I still sometimes heard the approving measures of John Bahcall in my ears. Yet those efforts also felt increasingly limiting to me. I had always been driven by the new, by the uncharted, by nosing my boat into an undisturbed lake, and that feeling had only grown with every new discovery, with every first. We weren't going to be surprised by a Hot Jupiter. We weren't going to find a life-form that can survive in flames. Surveying lifeless planet after lifeless planet . . . I could spend all the time I wanted in empty rooms in my own house.

A few years earlier, I had seen a British scientist named William Bains speak at an astrobiology conference in California. He'd intrigued me from the start of his talk. It wasn't his red hair and red beard that made him stand out in the room; it was his encyclopedic knowledge of biology and chemistry and the spaces where they interact. I loved the way he thought about life in the universe.

William brought his biotech class from England to visit Boston each spring. In 2009, he'd stopped by MIT and talked to me and my students about his research, which at the time involved finding liquids other than water that might sustain life. That led to expansive conversations about the different forms that life might take. Could life survive in jets of sulfur? Could life be based on silicon instead of carbon? That visit had gone so well, I invited him to stay in Cambridge for a couple of months to explore the limits of astrobiology with me. I am not a biochemist by training, but I wasn't about to concern myself with staying inside my field's arbitrary lines. The physical world defies the borders we draw across it. If I saw something worth exploring, especially if it would help me find other life in the universe, I wanted to pursue it.

We embarked on a brief, spectacular failure of a lab project, trying to cultivate a robust form of life on Earth—E. coli, in our case, the bacteria that finds our digestive tract a delightful habitat—in higher and higher temperatures. The biology professor who hosted us in her lab told us that our experiment was pointless: No, E. coli could not live on a planet like Mercury. She was right, it turned out. It was still fun to cook bacteria with William, playing mad scientists for a little while.

William became a friend, and, after Mike died, he asked me what he could do to help me with my grief. "Come visit MIT to work with me on biosignatures," I said. We reopened our efforts by pooling our imaginations. We talked about different gases that might be produced by life. We wondered what temperatures really did make life impossible, and what temperatures required us only to think differently about how life begins and survives. We knew science was just beginning to understand the incredible diversity of exoplanets: they come in every size and many colors; they orbit giant stars and dwarf stars and binary stars; they are made of combinations of solids, liquids, and gases. William and I made mental pictures of every conceivable world.

I was still especially interested in the composition of alien atmospheres. I still believed that Bigfoot's breath would give him away. When I wasn't thinking about new space telescopes to help us explore, I thought about what we should be looking for with them. William and I took a closer look at planetary chemistry, at different combinations of rock, surface temperature, and atmospheric mass, and how they might combine to alter alien skies. A planet's volcanism could also have a dramatic effect on its atmosphere. The highest mountain on Mars, Olympus Mons, is a volcano three times the height of Mount Everest. It's possible that other planets have volcanoes that are even larger than that, or more numerous or more active than our own. William and I reminded each other constantly: Anything is possible.

Our goal was to shake ourselves of our Earth-based biases—our "terracentrism," we called it, that peculiar blindness born of being human. Most of the scientists also at work on biosignature gases were using Earth as their model for other life-sustaining worlds. That's an understandable response to living on such a beautiful planet, just the right size and distance from its star. If we want to predict how much methane might be produced by life in a year, of course scientists would start with how much methane is produced by life *on Earth* in a year.

But William and I approached the math a different way. We knew that life-betraying gases are often destroyed by a cascade of atmospheric chemistry. That happens on Earth, too; the sun's ultraviolet rays smash other molecules into highly reactive components, called radicals, which in turn bond with all sorts of chemicals. William and I calculated how much of a particular gas needed to be present in an alien atmosphere for us to detect it with a future space telescope. Then we determined how much biomass would need to be present to make that much gas, factoring in the destructive powers of those same ultraviolet rays. If a planet would need trees ten miles tall to accumulate enough oxygen for us to see it, then we could eliminate it as a place we might look for life. Well, *probably* eliminate it. Maybe there is a planet with trees ten miles tall. Maybe there is a planet with trees that walk. Maybe there is a planet where the trees are kings and queens.

Next we asked another question: Can an atmosphere rich in hydrogen also betray signs of life? This was important to know, because hydrogen is a light gas. That means planets with a lot of hydrogen in their atmosphere look "puffier" than Earth does: Their atmospheres extend farther from their planet's surface than our delicate envelope. (Helium-filled balloons rise on Earth because our gravity is too weak to hold on to the helium molecules within them. The same would be true if we filled balloons with hydrogen. But a more massive or much colder planet could anchor hydrogen, so balloons would sink rather than float.) Puffier atmospheres are easier for us to detect—using the Transit Technique, we can more easily watch the spectrum of light passing through a puffy atmosphere than a thin one—which means that planets bathed in hydrogen would make for simpler candidates in our nascent search. But we weren't sure whether hydrogen might react harshly with biosignature gases, eating them up before we had a chance to detect them, the way Earth once consumed all of its oxygen. We used a computer to simulate every exoplanet atmosphere we could, trying to see whether their biosignature gases might survive hydrogen's grip.

They did.

•

I could feel my possibilities expanding, too—or maybe that's how it feels just before you break apart. I felt almost every emotion except happiness, and I felt all of them deeply. Some days I woke up with my pillow damp and saw no reason to go to work. I saw no reason to do anything. There were other times when I felt almost limitless, as though I'd built up an immunity to further harm. *What can possibly hurt me now?* I stopped worrying if the boys played recklessly, if they wanted to climb or swing from something they probably shouldn't. Let them try. *What's the worst that could happen? What could possibly compare?*

I told myself that I didn't owe anybody anything; I only owed myself a chance to smile. I'd think about Mike, and remember our first exploratory days together, and will myself into believing that there might be even better days ahead. *Someday I might even have a better best friend,* I'd think over and over, like a mantra. Then I would fall to the floor. Hour by hour, I felt either broken or bulletproof. I could span the space between them in seconds.

We had neighbors in Concord, an older couple, the Wheelers. I was walking home from the train station after a long day at work when they stopped me. "Oh, your kids are so sweet," Mrs. Wheeler said, which confused me, because she had never met my kids. Her husband had always been kind, but she had a schoolyard meanness in her, defining herself with her disdains rather than her affections. I braced myself for the inevitable turn. "They're leaving their toys in our space," she said. "And your yard is a mess. You put your fall leaves in a huge pile on our property. In Concord, we just don't do those things."

I exploded into sobs on the spot. Really? My yard? My husband died. It wasn't exactly a town secret. Leaves? A few toys? I couldn't muster a word in response; I felt lucky to find my next breath. She regarded my despair blankly. I recovered enough to return her stare only when I remembered that I didn't have to care about anything anymore, least of all what she thought of me. I walked away.

•

I moved out of the bedroom with the boys to the room across the hall. It was Max's room at one time, before he and Alex had decided to bunk together. When we first moved into the house, Mike had asked Max how he wanted his room painted. Max had asked for yellow walls, and then he'd asked Mike to add a giant rainbow to one of them. Mike had indulged his blue-eyed boy. I liked the rainbow. In that rainbow, I could see my life the way it once was. Because of some strange imbalance in our ancient radiator system, the rainbow room was also the warmest room in the house. I set up a little single bed in there, where I could curl up and feel like I was in the safest place.

I had a dream about Mike not long after I moved into the rainbow room. He hadn't died after all. He was back, as real as could be. He'd just been on a long canoe trip through the wilderness. It must have been a hard journey, physically demanding, because he looked beat. He was wearing shorts, a worn-through T-shirt, and a battered baseball cap. But he looked good, too, tanned and strong.

"Hi," I said, shocked to the point of speechlessness.

There were so many things to say, but my thoughts were tumbling, racing. For some reason, all I could manage to get out was "Mike, I threw out all your stuff."

He smiled. "It's okay," he said. "You didn't know." He was so gentle and practical about it. And then he disappeared. I woke up. My heart was beating fast. I looked at the rainbow on the wall and dissolved into tears.

Oh, Mike. You could be so kind.

After the last of his hope was gone, and he knew he was facing down death, Mike had made me a gift. He sat at his computer, pecking out what I would come to call my Guide to Life on Earth. Through his pain, he somehow managed to fill three double-spaced pages in the months before he died. Parts were filled with names and phone numbers. Who to call if this goes wrong. Who to call if that goes wrong. Plumbers and electricians. There were also notes on bills to pay. Which Montessori school to consider after the boys had outgrown their current one. He also wrote a list of chores, a catalog of reminders of what needs to be done and how to do them. Our house has a central vacuum. Of course the dirt it sucked up had to go somewhere, but I'd never given its fate a moment's thought. I didn't know that it went through a filter into a bag in a tank in our basement, and I didn't know that the bag needed to be changed before the vacuum stopped working. Mike's list included a reminder to empty that tank.

He must have sat there and wracked his fading memories

for the things he used to do, remembering the routines of his life before cancer, task after thankless task. Those three crumpled pages were my connection to his former world. They were a physical reminder of the man he had been and the things he did in his time on this planet. His literal and metaphorical DNA were all over them. My Guide to Life on Earth was a list of answers, but it also posed for me a single, overarching question: *How could I not have known?*

I had appreciated Mike when he was alive, but looking at that list, I wasn't sure I had appreciated him enough. Maybe I hadn't appreciated anything enough. Mike had given me the time and space to become an expert in the rest of the universe; he took care of home. Until he didn't. Now home was an arena for amateurs.

CHAPTER 11

Life on Earth

I could parent well enough. I kept the boys clean and well fed. They were always in bed on time. I answered their hard questions and set harder boundaries, even when it would have been easier to let them run free. They rarely had screen time; instead we spent hours in parks and at Walden Pond. I played tennis with Max and hiked with Alex. I even learned the special patience required to watch someone play LEGO just because they want you in the room.

It was more the physical world that defied me, the countless interactions with strangers and objects that guide us through the course of our days and weeks. I tried to remember the little girl I had been, forced to be brave by neglect. Once I had stood beside lakes in the dark and wintered in libraries and learned things nobody else had known. I could be that person again. I could do my research.

If Mike hadn't left me the answer to one of my questions, I sought it elsewhere. I suspect that I became the talk among the store owners in town: the lost, sad-looking lady who stumbled in and asked questions a child could answer. I actually came to enjoy some of the encounters. A new butcher shop opened across from the train station, and I decided that the two men behind the counter were going to help me learn how to cook. I went in, bought a steak, and asked for instructions. One of the men told me to cut it into thin strips and put it in a pan with a couple of inches of oil.

"That seems like a lot of oil. Okay, then what?"

"Then fry it on each side for two or three minutes."

"Okay, then what?"

"Then you eat it."

I trusted him, so I followed his oily-seeming instructions, and the boys and I sat down to a delicious steak.

Other lessons were harder to learn. It struck me as unnecessary that Mike shopped for groceries in four different stores, so I just took Max and Alex to Whole Foods. We had dinner at the Hot Bar, where Max wolfed down curry, Alex ate fruit and entire loaves of bread, and I'd pick away at chicken and rice. Then we shopped for groceries. A few months later, I realized why people call that place Whole Paycheck. I began following Mike's careful instructions more closely, buying staples where he had bought staples, produce where he had bought produce. I walked in his literal footsteps.

Groceries were the hardest, I think. Not just because I didn't really know what to buy or how to prepare whatever I lugged back home, but because there were always checkout lines, and lines gave me time to stop and think. My lonely commute on the Red Line was similarly flooded. I lost track of how often I stood in line and remembered Mike and felt the familiar rise of heat to my face. Nobody ever asked me what was wrong. People wanted to keep their distance from sadness, like it was contagious, like I had a disease. I would stand there by myself, turning into a puddle, worried that the bag boy would soon drop his head and get a mop: *Cleanup in aisle four . . .*

The employees of Rocky's Ace Hardware? Now they really did see me coming. They should have set up a system of alarms, a Rube Goldberg machine to lock their door the instant I darkened it. I might have had a growing army of help at home—Jessica got the boys ready for their day and babysat two evenings a week; Christine chipped in with mornings, cleaned the house, and sometimes left a dinner I could reheat; Diana was a constant during the after-school shift. But there were problems none of us could solve, for which I had to seek outside assistance. If I was walking into Rocky's, something had gone wrong at home, not procedurally, but with the house itself, which meant that I was either crying, had just been crying, or was about to cry. One of the men who worked there, whenever he saw me, said: "Ma'am, no need for the backstory. How can I help?"

In moments like that, I could see myself how others must have seen me: a crumpling woman, dressed in black, trapped in some kind of spiral.

One Friday evening I went to the library. Not long after we'd moved to Concord I'd joined a book club, a group of mothers from school. One of them had asked whether I wanted to come one evening. My initial response had been: *No, whatever for?* But reading the book seemed optional, and I was starting a new life in a new town. I was about to drive off to my first meeting when Mike had teased me. "You won't be able to make friends anyway. Why bother?" He said it with a smile. He knew me a little. He knew me better than anybody.

I had still gone. Now the old book club seemed like the easiest possible reintroduction to the outside world. Step one:

I had to get the book they were reading. I got off the train at Concord at 5:45 P.M. at the end of a long week. The library closed at 6:00, but I ran over and found the book. Prematurely triumphant, I headed for the checkout.

The woman behind the counter told me that I couldn't check out the book because I had a book overdue from another branch. I didn't know what she was talking about, but then I remembered that Alex and I had stopped by a distant branch in the spring to kill time while Max went to a friend's birthday party. That book had been long lost in the chaos of my house. I told the librarian there was no way I would ever find it. I would have to pay for it.

She frowned. A lost book. For a librarian, there is no greater sin.

"How much will it be?" I asked.

"Ten dollars."

I rummaged through my purse.

"You can't pay for it here," she said. She was still frowning.

"What do you mean?"

"You have to pay for it at the branch that's owed the book. The branch is so far away, it's part of a different system."

I just wanted to check out a book. I could feel everything rushing to the surface again. There was nothing I could do to stop it.

"If I can return it here, why can't I pay for it here? I don't know why you're singling me out. I don't know why you're trying to make my life so difficult."

She didn't answer.

"I work sixty hours a week. I can only come here on Friday night, in the fifteen minutes between my train getting back and when you close."

Now the other librarians were gathering around us. It was

maybe a minute before six. They just wanted to go home. We all just wanted to go home.

"My husband died. I have two kids. I work full-time. I just want to have a night out with my book club. BUT FIRST I NEED TO READ THE—"

"I'll take the ten dollars," she said.

•

My job was a magnet, one end or the other, the push or the pull. I found my sixty hours of work each week—maybe forty at MIT, and another twenty at home—either tiresome or therapeutic, depending on my mood that day. I didn't have a lot of patience for the drudgery of faculty meetings or campus niceties. If I was at a talk and it didn't interest me, I walked out in the middle. Time had become so precious, I refused to waste a second of it. But when I saw the value in my work, I still threw myself into it. When I saw how it mattered to the universe, it still mattered to me. My attention just wasn't automatic anymore—it had to be earned.

I worked on multiple projects at once, as I often had since grad school. They were each purposeful, each related to finding other life in the universe, but I liked toggling between different approaches to the same search. Both out of interest and strategy, I've always thought of research as a well-structured investment portfolio. I usually have some steady, safe work on the roster, conservative but with a fairly guaranteed yield, like my atmospheric research of real, known exoplanets with my students, including our studies of planet interiors and mini-Neptunes. I'll also dabble in something moderately risky but richer in possibility, like studying climate on rocky exoplanets—a challenge because it will be years before we can confirm today's theories with actual observations. And then I take a few really big swings, high in risk and reward. I take on more of that kind of thing than most researchers, and it's easily my favorite work. That's where biosignature gases fell in my mind. So, too, ASTERIA.

I'd made progress toward a prototype for my tiny satellites, inventing and testing precision-pointing hardware and software, and perfecting the design of the onboard telescope and its protective baffle. I worked hard to clear the rest of the path for ASTERIA to become real. After we'd laid the groundwork in the design-and-build class, my students and I were joined in our efforts by Draper Laboratory in Cambridge, where researchers work on things like missile guidance systems and submarine navigation. They also do a lot of work on space hardware. We had meetings every week, trying to solve the problems of small telescopes. We could build minuscule enough components, and we could deploy the satellite and tell it what to do, but we still couldn't figure out how to keep it as stable as we needed it to be. While we tried to solve that issue, I used my ongoing biosignature gas research to determine what types of exoplanets deserved our focus. I thought we might be able to explore a hundred star systems or so in my lifetime; they had to be the right ones. Stumped somewhere, I tried to make progress elsewhere. I never wanted to feel as though I'd reached a dead end. I wasn't sure I could stand it.

I helped teach the design-and-build class again. It remained a bright spot. David Miller, the lead professor, couldn't have been kinder. We weren't especially close, but he had been so supportive of ASTERIA, and he had a clear-eyed, levelheaded approach to everything. I arrived for the first class, the steep rows of empty red seats slowly filling with new students. David and I sat in the front row and waited. Nobody knew what to say when they were sitting in a quiet room with me. I didn't know what to say to David, either. But after a few moments, he turned and looked at me with a warm, disarming smile. "Welcome back," he said. It was the right thing to say.

This time the class would work on a new project, called REXIS. It was a proposed X-ray spectrometer, one of five remote-sensing instruments to be built for OSIRIS-REx, a NASA-funded mission. With David's leadership, MIT had won a NASA competition to guide a graduate-student build of REXIS, and $5 million to get the job done.

Some objects in space require us to go to them. Some are coming to us. OSIRIS-REx was a planned mission to visit an asteroid now named 101955 Bennu. (It was called 1999 RQ36 at the time.) 101955 Bennu is what we call a "carbonaceous asteroid"—the most common kind, made mostly of carbon— about five hundred meters in diameter. There is a faint possibility that 101955 Bennu will slam into the Earth sometime late in the twenty-second century.

There is an algorithmic scale, the Palermo Technical Impact Hazard Scale, that classifies near-Earth objects in terms of the risk they pose to our planet. 101955 Bennu is in the second-highest risk category, with a 1-in-2,700 chance of colliding with us. It's about one-third the size of the meteor that carved out Chesapeake Bay, so we should probably try to avoid it. (The bomb dropped on Hiroshima had a blast yield of 15 kilotons of TNT; 101955 Bennu would be the equivalent of setting off about 1,200 *mega*tons. There are 1,000 kilotons in a megaton, and it's better not to finish working out the math.) OSIRIS-REx would visit the asteroid and return a sample of it to Earth. Our component, REXIS, would look a little like an advanced and possibly overengineered microwave covered in gold foil. It would hitch a ride on OSIRIS-REx, land on the surface of 101955 Bennu, and take an X-ray of the entire asteroid. That would allow NASA's scientists to con-firm that the sample OSIRIS-REx brought back to Earth was representative of the whole.

The entire mission was built on the premise that it's better to know your enemy. It wasn't something I'd normally help tackle—it had nothing to do with finding another Earth; it was about saving this one—but I was happy to work with my students on such a practical and necessary project. The hard-ware store might overwhelm me, but space hardware I could understand.

•

Can you imagine someone with much more sophisticated tele-scopes than ours, looking for us and seeing the orange glow of our cities at night? The lightless rectangle of Central Park? The black ribbon of the Seine finding its way through Paris? The reverie made my heart beat faster. The reality, however, was that so much of my work was still just math: theoretical, statistical, sensing not seeing. Numbers were too often all we had.

When Mike died, our family of four became a family of three. Mathematically speaking, his departure represented a significant loss: 25 percent of our household. More critically, however, we went from an even-numbered family to an odd-numbered family. That might not seem like it should matter, but in the way our world can seem custom-built to defy left-handed people, we also live under the tyranny of even num-bers. Deep down, a lot of people find something imperfect or unsatisfying in odd numbers; they're like building kits that come one part short or with one part too many. The arche-typal nuclear family, the mythical bedrock of American soci-ety, is two adults and two children. We see in those numbers

balance and symmetry, a square root and division without re-
mainders, and we have built our society on the foundation of
that mathematical ideal.

Everywhere Max, Alex, and I went, I was reminded that we
were incomplete. Not just by our own accounts, but by our
suburban society's greater calculus. If you're an adult, it's almost always assumed that you are part of a couple. Or, if
you're not part of a couple, then you aspire to be. A table for
one in a restaurant always has a second empty chair beside it,
just in case you need the confirmation that you are missing
someone. And if you're a family, the working assumption is
that you're a family of four. Cars and restaurant booths, roller
coasters and family tickets to the museum: Two adults and
two children. Two plus two. Two by two. Four.

I forced myself to go to work-related social events, just to
get out of the house. That meant leaving the boys with Jessica. It was too soon. I almost always regretted it. Once I went
to a big dinner with a guest speaker at work, unaware that
most of my colleagues were bringing their partners. Someone
said to me: "Well, you're single, so you sit here." What I'd
once found so comforting in MIT—the eminence of logic, of
bluntness, of practicality—now sometimes hurt me. Simple
statements of fact had never been so cutting.

Humans are bad at dealing with damaged humans. If people said anything to me about Mike, they almost always said
the wrong thing. *I have no idea what I'd do if my husband
died*. That's not the best message for a widow to hear, but
widows hear it all the time. I learned to break the news of
Mike's death slowly to people who didn't know, as though
they were the ones who needed protection, not me. "You know
Mike was sick," I'd start. Nod. "You know the chemo wasn't
working." Nod. I'd keep going until their face registered the

appropriate discomfort. Others would ask about Mike or "your husband" in more generic terms, and I learned that it was better to deflect those questions than answer them directly. "Why doesn't your husband do more to help out?" or "What does your husband do for work?" Those were hard. I coached the boys on how to answer similar questions from kids at school or summer camp. We'd practice glib, vague responses, and build them into a sort of arsenal, a collection of the darkest inside jokes. "Well, he isn't able to talk right now" or "Gee, he's really busy at the moment" or "He's away on a long trip." The three of us would laugh until our laughter turned into something else.

I went to a colleague's wedding, and that was a form of self-harm. After he and his new wife exchanged their vows in a solemn Eastern Orthodox ceremony, I was the only single person left in the room. Then came a loud, rousing party. Everyone was having fun, drinking together and laughing together and dancing together. I left alone. I walked back to my car, found a parking ticket on the windshield, and drove back to my single bed through the first heavy snow of the winter. I didn't know what grief triggers were then, but Eastern Orthodox weddings must be high on the list.

What I've always loved about numbers had become something for me to resent about them. Numbers are black and white, binary. *Numbers don't lie.* Now they were a reminder of my family's new, permanent incompleteness, condemned to be odd.

•

Traveling for work, which I now did again for short trips, made me feel most alone: I ate far too many dinners for one. For a couple of days, while Jessica looked after the boys, I headed down to the Space Telescope Science Institute in Baltimore, home of Hubble's science operations. I was on the advisory committee for the James Webb Space Telescope, the more powerful version of Hubble that would operate in the infrared.

On that occasion, at least, I didn't have to eat alone. I went for dinner with Bob Williams, the same Bob Williams who had defied so many of his peers to explore the Hubble Deep Field. I needed the same inspiration I had found in him before. We met in a wood-paneled room in what he called the best brewpub in the city, the after-work crowd just starting to build around us. We sat at a little square table and ordered drinks, then dinner. I asked him to tell me the story of the Hubble Deep Field again, those ten days of dissent and discovery. Bob sat back and smiled. He's lean and confident, a natural athlete and raconteur, and he didn't need a lot of convincing to narrate his greatest triumph. In his soft Southern accent, he recounted how he had decided that it didn't matter to him what anybody else thought. He had earned his rights to Hubble. "I was going to look where I wanted to look," he said. Hubble turned at his direction toward that pitch-black patch of sky. And Bob found those three thousand new galaxies. He'd found those billions of new lights.

This time, the story of the Hubble Deep Field did not have the desired effect. I don't think I'd ever felt lower. I sat across the table from one of history's great explorers, while he regaled me with an astrophysical epic, and all I could do was fight not to cry the wrong kind of tears.

Poor Bob. This kind, wise man, who was helping to solve so many mysteries—he had no idea what to do with a weeping widow. The depths of the universe didn't flummox him; the depths of my depression did. He tried to engage. He knew,

I think, that I looked at him as another of my father figures, and he did his best to give me what I needed. But our dinner was like a torturous game of tennis. He'd come up with a suggestion, something that I could do to make myself feel better—therapy, meditation, some time away—and I'd shake my head, knowing that I'd already tried it or that it wouldn't work.

"Sara," Bob said finally. "Do you know what I do when I need clarity? I run across the Grand Canyon. In a single day."

He knew that I loved the outdoors, loved movement, loved physical accomplishment, loved tangible goals. He didn't know that Mike and I had gone to the Grand Canyon together all those years before, in a different life. *The Grand Canyon?* I remembered us on the water that magical day, when Mike had basked in the cheering of strangers. It was a world away from where I was, but thinking about it seemed like the right kind of expansion.

I stopped crying.

•

Max, Alex, and I initially skipped holidays, ignoring their existence and whatever memories, good or bad, they might bring to the surface. They had never been important to Mike and me, and we weren't close enough to our families to feel pressure to celebrate arbitrary days on the calendar. Now they seemed more risk than the little reward I thought they were worth. I had begun inviting over groups of students and postdocs—Alex orchestrating an egg-eating contest that got a little *Cool Hand Luke*—and once I had attempted an imitation of Thanksgiving for them between actual Thanksgiving and Christmas. That was a mistake. I couldn't get the turkey cooked properly, the legs defying me, and then I remembered that whenever we had eaten turkey as a family of four, Mike would eat one leg and Max would eat the other. I lost it.

I decided to take the boys and Jessica to Hawaii for our non-Christmas. Jessica's family celebrated Christmas on Christmas Eve, so we would leave on Christmas Day. An airplane over the ocean was the best isolation chamber I could imagine. It wouldn't feel like Christmas up in the air.

Alex hadn't lost his love of hiking. I had mixed feelings about it, honestly—for paddlers, a hike feels like a portage without water at the end of it—but I wanted to have a shared adventure and to help him do whatever it was he decided to do. I told him that there was a mountain in Hawaii that we could climb: Mount Haleakala. It's a massive shield volcano, the beating heart of Maui, and it would be a long, 10,000-foot hike to its crater.

I was a little worried that Alex wouldn't be able to manage it. Before the trip, I was at a conference where I bumped into a colleague from the University of Hawaii in Oahu, and asked him about the hike. He scolded me: "Haleakala is no place for a child!" I told him that I was on a mission, that I was trying to empower my kids, to make them feel as though the world and their places in it could still be beautiful, even if I didn't always feel that way myself. I told him that Alex had climbed mountains before and that he was special when it came to determination. "Everybody's child is special," he said with a scoff.

I was undeterred, but I decided we should hire a guide, someone who knew the route, just to be safe. In the back of my mind I thought that if worse came to worst, a guide could help me carry Alex out of trouble. I found a company online and called to see if they would help me and a six-year-old boy climb the mountain. The man had owned the company for decades. I explained the situation. I had spent a lot of time outdoors and knew my way around long journeys, and Alex wanted to set world records. I was sure the man was going to say no, but he agreed to help.

We woke up at four-thirty in the morning to begin our quest. I couldn't find a coffee to save my life, which might have set me off, but I held it together. Jessica and Max drove us, along with our guide, Dylan, to the start of the trail. I told them to enjoy their day and meet us at the top at six that night—a road wound up the other side of the mountain—but they should bring some books and blankets in case we were late. We said our goodbyes, and Alex, Dylan, and I started our hike in the chill of the dawn, under the forest canopy. We would follow the Kaupo Gap all the way to the summit.

It was a long, long climb. We were lucky in a lot of ways. The morning stayed cool while we climbed the first 7,000 feet of elevation. When we walked into the expanse of the crater floor, a cloud blocked the burning sun like an eclipse. Later, a mist of rain came and cooled us off. We saw a rare, stunning plant unique to Haleakala: the silversword, a succulent covered in silver hairs. But we still had to cover a lot of ground. After fifteen miles, we reached the last switchback to the summit—another three miles up what was essentially a steep dirt road. Alex was tired, and his shins had started to hurt. I told him that Dylan could carry him if he needed to be carried. It was up to him. He said that he wanted to finish the climb.

Night fell. Orion, Alex's namesake constellation, came out. On we climbed. We were a little late reaching the top—it had taken us more than thirteen hours of solid effort in the end—but we made it. Jessica and Max were waiting in the car. We fell into it and drove back down the mountain, back to our hotel. Alex made a little sleeping nest with blankets on the floor, and he lay down and fell asleep. The rest of us went to get dinner at the hotel restaurant, and when we came back, Alex still hadn't moved. I shook him gently. "Just say your name," I said. He didn't make a sound.

I fell asleep, too. When I woke up the next morning, Alex was sitting beside me with a huge smile on his face. In a small but triumphant voice, he said: "Mom, you said if I made it up, you'd take me to climb a 14,000-footer. I made it up."

I felt a moment of buoyancy, an almost forgotten rush of good feeling. I didn't know how we had made it, but we did.

By the time we had packed and driven back to our rental apartment on the other side of the island, the feeling had mostly evaporated. I had gone dark again, tired and sullen. My food felt and tasted like sand in my mouth. I realized that whatever happiness I might feel anymore was like an island rising out of an ocean: My elation was Maui, my despair the Pacific. My joy was Orion, surrounded by the night sky.

We flew back to Concord, back to all the reminders and memories, back to the depths of literal winter. The cold and the dark and the routines of a snowbound life. I felt worse than I did before our trip, the way I imagine someone who escapes from prison must feel after they've been captured and returned to their cell. I had seen a sliver of a different, better world, but I wasn't allowed to stay there.

It wasn't long before I had another dream about Mike. He came back to me again. He always started the dreams standing outside the house. He stepped into the foyer. This time he didn't look as rugged, but he still looked good. He wasn't returning from a trip. He'd been sick; he'd been in a coma. If he had been awake, he told me, he would have let me know, but he'd been asleep.

In my dream, I felt ecstatic at his return but also bewildered by it. His coma story made some sense, or at least dream sense, but it also seemed unreal, like the storyline of a soap opera. I looked at him and smiled before I remembered that I had to deliver the same blow that I had given him in my first dream: "Mike," I said, in my sweetest voice. "I didn't know you were coming back. I got rid of all your stuff."

This time, he didn't say that he understood. This time, he was mad. "What?" he said, his voice lined with a rising anger. How could I have been so cold? I woke up with a start. I woke up in tears. I woke up alone in my bed, adrift in the middle of the ocean.

.

A week or two later, I began having serious pain in my lower abdomen, cramps that could make it hard for me to function. I had spent more than enough time in the company of doctors, diagnosing stomachaches, but I made an appointment. Widows often experience physical pain, and too many doctors are quick to dismiss their aches as psychological. I remembered Mike's ruined ankle and pushed my doctor to authorize tests. She sent me for a uterine ultrasound.

I sat in the waiting room with a dozen expectant mothers and their nervous, happy husbands. Their full bellies, their anticipation, their blessings and happy news . . . Being surrounded by so much life nearly killed me. I staggered into the ultrasound room and flinched at the cold of the jelly and the pressure of the wand on my skin. I lay there and remembered what my life used to be. The last time I'd felt that wand, I was watching Alex's heart beat. This time, the ultrasound didn't find anything unusual for a woman over forty. I went home afterward, barely made it through the numbing evening routine, and collapsed into bed.

The next morning, it was only the distant laughter of my boys that persuaded me to push back the covers. After a quick breakfast, Max and Alex began putting on their snowsuits. With their plastic sleds stuffed into the car, we made the short drive to the top of Nashawtuc Hill.

- - - - - - - - - - - - - - -

CHAPTER 12

The Widows of Concord

Minnie May died not long before Valentine's Day. She had outlived Mike after all, by more than six months. It wasn't that close of a race in the end. I have no idea how she lived for as long as she did; she looked like a bag of bones near the finish. People gasped when they saw her. For more than a decade, she'd had medication to ward off epileptic seizures and bladder stones, and was even on Prozac for anxiety. She'd had more than her nine lives. But her heart still beat, a tiny engine custom-built to resist the finality of death.

One day Minnie May's back legs stopped working. Some invisible internal switch had been flipped. She didn't seem to be in pain. She didn't seem to feel much of anything. I carried her up to my room and wrapped her in blankets. That night I woke up every hour to put my hands on her and feel her shallow breathing. I woke up one last time and reached out to find her in the dark. Minnie May had always been still something: still breathing, still living. Now she was only still.

She had been there for eighteen years, for every significant moment of my adult life, good and bad. She was my most faithful observer, and she could make me feel as though a chosen few of us might live forever. Her death felt like the hardest scientific proof that none of us do. If Minnie May could die, then everything will.

The hole that Mike had dug for her in the yard was full of snow; the pile of waiting earth was hardened into clumps. Despite his best efforts, I had to put Minnie May in the basement freezer, my makeshift kitty morgue, just as I'd stored Molly until the thaw. I came back upstairs and sat down at the kitchen table in the quiet of my once full-to-bursting home. There was a time in my life when I had my father, my husband, my dog, and my cats. I was surrounded by life and love. So much had fallen away, like parts coming off a machine. Even Jessica had decided to move out in early 2012, amicably but completely, leaving behind only my boys and me. My world was stripped down to its glowing core.

A few days later it was Valentine's Day. Melissa, the widow from the hill, had told me on the phone that the boys were also invited. Max, Alex, and I were going to a party.

•

Equal parts nervous and hopeful—maybe more nervous than hopeful—I left work early to plan what to wear. I wore a black shirt, but I didn't want to wear only black to the party. I wasn't in the mood for romance, but I didn't want it to feel like a funeral either. I wanted to dress for a celebration, even if I wasn't sure of what, exactly. I pulled on a calf-length, fawn-colored cardigan and a bright pink scarf. I had gone for a manicure earlier in the week, and my pink nail polish completed the outfit. The flashes of color made a surprising psychological difference to me. They were something new in my black-and-gray world.

I'd bought some heart-shaped pasta to bring. I was delighted when I'd found it in the butcher shop. There was an older man doing his shopping not far from me. He worked at MIT. At the time, the government classified all satellites as weapons—in the wrong hands, they could be—and there were rules about how and when foreigners could work on them. This man's job was to make sure we didn't break those rules with international students and visitors. I'd met with him to discuss ASTERIA a number of times, and we shared the same train. I thought of him as a friendly face.

"I'm so happy about this pasta," I told him. "I'm going to this Valentine's Day party for widows . . ." He looked aghast, as though I'd let slip my plans to launch an actual weapon into space. He must not have known I was a widow, even though I figured everyone did. He took a step back as though he were afraid of me. I had always unsettled some people, but even when I was a little girl, people weren't usually so obvious about their discomfort. That man looked at me and seemed to see someone in violation of the rules.

Melissa told me that there would be all six of us, including me, and eleven tragically united children. We all lived within a couple of miles of each other, which still struck me as an unlikely coincidence. Our otherwise quaint little town was under a terrible jinx. The boys were a little nervous, since we didn't often go to social events that weren't related to my work, but I had done the math. "You will probably know at least one of the kids," I said. Between soccer and camp, the chances were good.

The house was giant, with a bright and cheery kitchen. It belonged to Gail. She was a little older than me and a commanding presence in the room, maybe because she was the only one of us in familiar surroundings. The other Widows had met as a group only once before, and now they were becoming reacquainted. We gathered around Gail like students drawn to the teacher on the first day of school, leaning against the counter, keeping our hands busy with our drinks when we weren't extending them to one another. When dinner was ready, we worked together to help serve the meal, another plate of food for another Widow.

There was Pam, the youngest of us, fashionably dressed and in tremendous shape. I was struck by how perfectly straight her hair and teeth both were. It took me a second to place Micah, but we had met once before: We had talked to each other in the park after she'd asked me about my jade necklace, a gift from my father. I remembered, too, that her husband had been with her that day. I couldn't help wondering what awful fate had befallen them after that. A short dark blonde named Diane stood quiet in a corner, looking as shy as I felt. And then there was Melissa, smiling and radiant, even more beautiful than I remembered her at the top of Nashawtuc Hill. Her red hair and perfect pale skin made her look like a flame.

The boys disappeared. I felt a bit lost, standing there without them, trying to make small talk with people I knew only because something had happened to each of our husbands. I listened more than I spoke. I watched more than I performed. The Widows all looked healthy. They all gave off a similar glow. I wondered what they thought when they looked at me, whether they saw a star or a shadow. I didn't usually wonder what people thought of me; I assumed a measure of distance, though I was still years from the diagnosis that would help me understand why. But now I was at a party for women who were, in at least one towering respect, just like me. After about ten minutes, Max and Alex reappeared in the kitchen, practically floating. They wanted me to know that I had been right after all: One of the kids was a boy their age they knew from camp. Then they were off again. Their happiness was infectious, and I felt my guard begin to drop. I could be something like myself.

One of the weird things about meeting the Widows: None of the other women was working outside the home—their husbands had been professionals, or they had run businesses together that had been sold after their deaths—so nobody asked me about my work. I was relieved in a way. I had felt anxious and tentative enough, being new to the group. Before I'd walked through the door, I had tried to rehearse how I would talk about my work with them, and in my mind the conversations had never ended well:

"So, Sara, what do you do?"

"I teach at MIT."

"What do you teach?"

"Planetary physics."

"Wow. Um . . . What?"

"I'm looking for planets outside our solar system. Other stars presumably have planets. I'm looking for them."

"Why?"

"Well, I'd like to find other life in the universe."

"You mean aliens? You're looking for aliens?"

"Scientists don't call them aliens. Other life."

"Right. So . . . Aliens?"

At least for now, it was better that the Widows knew me only as a fellow luckless one. In that fundamental aspect we were identical. At their previous, first meeting, the Widows must have stopped short of true revelation, sticking with the usual chatter about kids and schools and hometowns. Now they began sharing the details of their lives that everybody really wanted to know.

"How did your husband die?" someone asked. "When?" "Cancer," I said.

Two of the others had lost their husbands to cancer, too. Melissa's had died bicycling. Riding down a big hill, he had struck a mouse skittering across the road, gone over the handlebars, hit his head, and died on the spot. *Instantly*. Another had died hiking. One had committed suicide.

Next we did the dates. "Death Days," we would call them.

"July twenty-third," I said. Mike had died nearly seven months before. In a little more than five months, I'd wake up on the first anniversary. I was lost in my own thoughts again. I saw someone writing down our names and the dates, but I couldn't imagine why.

We retreated to the dining room for dinner, each taking our place around a huge, formally set table. Gail sat at the head of it, and the rest of us buzzed around her, excited about the possibilities of knowing one another, and about the night's escape, however brief, from our usual slogs. Suddenly, Micah looked at Gail and blurted: "So, Gail, are you dating?" The rest of the conversation stopped. We all looked to Gail.

"No," she said. "I went on one date. It didn't go very well." I felt as awkward as I had at the start of the night. I wondered if we were going to start prying into each other's private lives, if I'd have to take my turn sharing fears and secrets. I wasn't ready for that.

"We signed a ketubah, a Jewish marriage contract," Gail said. She pointed to a beautiful framed print that was hanging on the wall, small illustrations decorated with the most delicate calligraphy. "The ketubah ends with *from now until forever*." I recalled my own vows: I had agreed to be married to Mike *till death do us part*. I was just beginning to process the difference between them when Gail's wineglass shattered. It exploded in her hand. There was a round of nervous laughter, and the conversation changed.

We talked into the evening, tying invisible strings to one another with every shared experience, with each new admission. Gail's father was there—he had been made a widower recently, if more expectedly—and he took a shaky picture of us standing shoulder to shoulder and already a little braver for the connection. MIT had given me a sense of belonging, that feeling of finally being around some people sort of like me, but this was different. Widowhood had made it seem impossible that I would feel anything more than a partial affinity with anyone again. But in Gail's house, I wasn't alone, one of one. I wasn't *her* or *them*. I was *Sara*. I was *nice to meet you*. I was *us*.

It felt really good to be able to use that word again: *Us*. It felt like a warm, bright light.

●

The Widows decided that we would get together for coffee every other Friday morning, moving from house to house like emotional squatters, filling someone else's too-quiet home for a little while before moving on to the next proxy for our own grief. Most of our children were old enough to be in school. We could share our feelings knowing that only the right sets of ears were hearing them.

Only a few weeks after Valentine's Day, it was my turn to host. It was a shining March morning. Light streamed through the big bay window in my living room. I sat and waited for everyone and reminded myself how important it was to listen. Three or four of the Widows arrived, and Gail, always generous, brought me flowers. Remarkably, a new widow would be joining us. There was a large potted plant awaiting the newest member of our club. We would all be listening, it turned out.

Her name was Chris. She was from the neighboring town of Lexington. We'd heard that her husband had died only a month before, in February, skiing. He'd hit a tree. That winter had been warm, and there hadn't been a lot of snow. I was surprised they had been skiing at all.

But the strange weather, the lack of snow . . . It wasn't hard for us to imagine what had gone wrong. We all knew how little it took for lives to change. Maybe the conditions weren't great. Maybe there had been some freezing and thawing. That would have left exposed earth in the wells around the trees, and patches on the hill with more ice than snow. Chris's husband no doubt threaded his way down his chosen slope the way he had thousands of times. He probably reached his usual speed, but maybe he couldn't find his usual control. He lost an edge and sailed into the trees, flying headlong into a trunk, crashing into a heap at its base. He had been filled with life, and then he had taken the wrong path down the wrong slope on the wrong weekend, and just like that, he was gone. His wife would spend the rest of her life hearing that at least her husband died doing what he loved. I knew it would never give her comfort.

When Chris joined us, she made seven. She was about my age, and her children, a boy and a girl, were about the same ages as Max and Alex; they made thirteen. Chris stood for a long time before she sat with us in the living room. When I first took her in, I thought she was doing really well. She had taken grief leave from her job as a data analyst. She was dressed head to toe in black, but her hair and makeup looked great. Then she began crying the instant she tried to speak.

She sat in a soggy pile on my couch, her tears falling into her coffee. She asked us to excuse her sadness; such apologies were always needless. She managed to tell us the date of her husband's death and not much else. The when. That's all, really. But her exorcism had begun.

There were patterns in how we shared our stories. Information came in rushes, with weeks- or months-long doldrums of private calculations in between. I was never sure whether the breaks were for the benefit of the teller or the listeners— whether that time was for the teller to recover her strength or the listeners to digest the facts and figures of some new agony. It felt as though we shared a burden whose combined sum never changed; only its distribution did, and we kept a sense of unspoken balance between us. If one of us talked, if one of us unloaded some of our pain, then next we had to help carry someone else's terrible weight. We were a kind of mule train, walking in our crooked line over impossible terrain, each taking our turn in the lead, each giving the others a spell from the heaviest packs. We knew Chris's story, like all of our stories, would come out later, maybe in pieces, maybe in a burst. We would nod knowingly whenever it did. None of our stories was any worse than any of the others, nor were any of the impact craters. They all had the same end. Seven dead husbands had left behind a single mourning wife.

Chris recovered enough to stop crying for a moment. Her red eyes took in each of us, my living room filled with light, fresh flowers, and plants in their tidy pots.

I hoped she could manage to care for hers. I kept my thoughts to myself, but I wondered whether Chris was ready for us. I wouldn't have been when I was where she was. A month after Mike's death, I was still in the middle of my feelings of liberation. I hadn't yet lost my way trying to get a haircut. I hadn't yet learned from Freya about the fall that comes after the fall, or begun meeting the agonizing demands of wills and the world, or found the complicated company of men in hardware stores and butcher shops. Chris was still taking her kids skiing every weekend. Packing them into the car and driving to the mountains of Vermont. It was exhausting, but it was what they had always done, and she hadn't yet learned what else to do. Five or six months removed—that was the best time to join our makeshift group therapy sessions, I thought: after a few introductory lessons in the hardest truths. But it wasn't like we would excommunicate Chris. We couldn't tell her that it was too soon. She had become one of us the day her husband had taken his last chairlift ride up the hill.

"Oh God, I'm such a mess," she said, "and you're all so put-together. Sara, your house is so clean."

I caught my laughter before it rushed out of my throat. Chris was looking at us the way I'd looked at Melissa that morning on the hill. It was with a kind of envy, the sufferer facing the survivors, the newly stricken looking across the gulf at those who had made it to the other side. She had no idea that I'd spent the last several weeks trying and failing to find the strength to pick up Mike's remains from the funeral home. Dave, the funeral director, had told me time and again not to worry about it. "Mike doesn't care how long it takes," he said.

One day later that spring, I genuinely thought I was ready. The warmth in the air hit me as soon as I stepped off the train. I walked across the street and was already on the verge of tears when I walked through the door. Dave greeted me in his usual, practiced way: cheery, but not unacceptably so. He invited me into his office. I couldn't really speak. "I'm going to cry," I said, and then I did. Dave smiled. He had the most disarming way about him. Funeral directors might be the best readers of human need on the planet. Dave knew that I didn't want his pity, and his smile wasn't a pitiable one. He looked almost amused. "I knew you weren't ready, Sara," he said. "He can stay with us for as long as you need." Dave smiled again. "Mike has plenty of company."

•

Our Friday-morning coffees got steadily lighter in tone. They became less about the deaths of our husbands and more about the new lives we were learning to lead. There were only so many times and so many ways you could rail at the same fate. I found the pooling of practical information hugely valuable. I was still learning how the world worked, and the Widows were so wise. They each wrote more lines in my Guide to Life on Earth.

I had never cared about money, and after Mike's death I cared about it even less. I had always lived beneath my means, so I never really worried about it. Besides, it had always struck me as such a strange, arbitrary invention. I don't remember agreeing to subscribe to the idea that the same paper is worth different amounts because it was stamped with a different number. It's ridiculous. It's like how we decided—or someone decided—that diamonds are worth something. They're shiny rocks. Quartz is a shiny rock. So is coal. Why should I accept that diamonds are worth more than them? Mike's frugality had made its impression on me, as had my mother's deprivations. I understood that it was better to have money than not. But that was the extent of my thinking. I had no idea how much was enough, or how to acquire more of it, or what to do when I had it.

The other Widows thought harder about money. They talked about it frequently. I realized that I had better try to understand its intricacies, especially how to make it last. We talked about the fair wage for a babysitter and what we could claim on our taxes. We talked about Social Security benefits from the deceased, and how much we could set aside for each of our fatherless children.

We talked about men only slightly less than we talked about money. Maybe half of the Widows had started dating, and each Friday we received a fresh roundup of that week's horror stories. One week Micah reported that her last date had been so bad, she wished she had stayed home and sorted out her kitchen cupboards instead. I didn't really participate in those conversations. I didn't like talking about men or how badly I did or did not need one. I was far from even thinking about finding another love.

I did feel comfortable enough to start sharing a little more about my work. My department head had given me the spring semester off from teaching, although I went into the office most days for meetings and research. The stars were such a fundamental part of my existence that not talking about them felt somehow dishonest. Still, I was careful not to get too far into the weeds. My most recent work focused mostly on bio-signature gases and ASTERIA, slightly esoteric subjects for chitchat; but my students and postdocs worked on broader, more relatable research, including Kepler's frequent news-making finds.

The search for alien life was feeling less and less like science fiction—or, worse, like the purview of conspiracy theorists and homebound weirdos—and more like science. I had sometimes felt stung by the mocking dismissal of skeptics, by people who ignored my credentials and the odds against our being alone in the universe. I cringed when people, usually politicians looking to cut budgets, made jokes about Area 51 or anal probes or the latest rendering of a bug-eyed alien in *The National Enquirer.* I tried to help people see that it's stranger to think we're the only blue light in the sky. There are billions and billions of planets out there, I'd point out. Statistically speaking, what are the chances that ours is the only one that sustains life?

That previous December, NASA announced with great fanfare that Kepler had found its first small planet orbiting in the Goldilocks zone of a sun-like star: Kepler-22b. We still didn't know much about our distant neighbor. All we had was its size and its orbit, but that was enough to garner a lot of attention. Kepler-22b is only a little more than twice the size of Earth, and at the time it was the smallest planet ever found inside another star's Goldilocks zone.

Some things got lost in translation between the scientific community and the mainstream press. There were too many headlines trumpeting the discovery of ANOTHER EARTH, when Kepler-22b is not nearly as hospitably constructed as our planet. A planet its size most likely has an atmosphere much thicker than ours, which probably means that it is surrounded by a suffocating envelope of greenhouse gases. Or perhaps Kepler-22b's atmosphere is so deep that the planet doesn't have a solid surface in the way we think of a solid surface. Either way, it almost certainly doesn't sustain life.

The discovery was still critical. In only a decade or two, we had made giant leaps in our planet-finding abilities. Our instruments were allowing us to see smaller objects closer to their stars, and smaller is good. Smaller probably means thin-

ner atmospheres, and cooler temperatures, and rocky sur-
faces. Planets like Earth must be out there—perhaps millions
of them. One day, I told the Widows, I felt certain that "we'll
know we're not alone." They looked at me and nodded politely, supportive as always. And then we began talking again
about how to survive the absence of our men.

CHAPTER 13

Stars Like Pearls

We huddled together on a big patch of concrete at an old missile site. Night fell, desert-hard and blacker than black. It was a little unnerving, being out there. Without a sliver of moon, the stars were unchallenged and bright, making a canopy of the purest white light. We looked at them as though seeing them for the first time.

I was in the middle of New Mexico, to test out a new component for ASTERIA. I was more and more certain of its value. It wasn't Hubble or Spitzer or Kepler, and it might never be something so magnificent. But not every painting should or could be *Starry Night*. There is room in the universe for smaller work, a different kind of art. Kepler might find thousands of new worlds, but it wouldn't reveal enough of any single one of them for us to *know* whether it was somebody's home. It was sweeping its eye across star fields that were too far away for astronomers to make anything more than assumptions about places like Kepler-22b.

But if I could just make ASTERIA work, and then find a way to send up a fleet of satellites . . . It would combine the best outcomes of Kepler, capable of finding smaller planets around sun-like stars, and the nascent TESS, with its more proximate search and sensitivity to red dwarf stars. I had dropped out of the TESS working group when Mike was sick, and REXIS had been shipped out of our design-and-build class. Now ASTERIA was my favorite machine.

My team built a prototype for a possible camera, one that was both promisingly stable and could operate at a warmer temperature than the detectors used in most satellites. (Most detectors have to be cooled, which taxes the machine.) I just wasn't sure that it would see what we needed it to see. I had a particularly bright and enthusiastic grad student at the time, named Mary Knapp; she had been an undergraduate student in the first design-and-build class I taught. She encouraged us to test the camera outside, using it to look at real stars. Mary proposed the deserts of New Mexico as our proving ground. That April, there would be a new moon, casting the already clear desert sky an even pitcher black. That new moon also coincided with Max and Alex's school break, which meant that I could take them along. As much as I wanted to see the stars, I wanted to see them, too.

We rounded out our team. Along with Mary, we'd be joined by Becky, a research assistant who had also been in the ASTERIA design-and-build class; Brice, the Swiss postdoc; and another postdoc named Vlada. (Vlada also happened to be magnetic, dark, and gorgeous, another Swiss, of Serbian descent. The Widows would have been all over him.) I thought of the trip as a vacation as much as work. I'd be doing something hopeful with a group of young, energetic people, in a place far removed from the trials of New England. It was another possible taste of my new, next life.

I was given a quick lesson in how pessimistic my default view of the world was. The gang met Max, Alex, and me at the airport in Roswell, picking us up in an enormous SUV, and they erupted in congratulations the instant we entered Arrivals: Word had gone out that I'd been awarded the Sackler International Prize in Physics. (It's an award given to young scientists who have made some significant, original contribution to their field. I received it for my work on exoplanet atmospheres.) I was honored, of course, and the award came with $50,000, which I was more than happy to accept. The catch—because my life could sometimes feel like a long series of catches—was that I would have to go to Tel Aviv University to receive it. It would be an eleven-hour flight each way for a two-day stay, so taking the boys with me was out of the question. I would be forced to leave them at home, something I normally did only for short trips, close to Concord. I would be farther away from them than I had ever been, and the distance scared me. I panicked thinking about leaving them without an immediate family member. What if they got sick? What if one of them broke his leg? Instead of being excited about New Mexico, I fixated on what was, to my mind, a serious problem.

We drove from the airport to the hotel. I stared out the window and tried to lose myself in the endless variety of the desert. Sometimes New Mexico looks like the moon; sometimes it looks like Mars. I was thinking that maybe somewhere on Earth resembles every rocky planet, every moonscape, when one of the boys announced that he needed, very badly, to use the toilet. We were on an empty highway in the middle of nowhere. The rust-brown horizon was broken only by parched bushes and the occasional strong-armed cactus. He was shy without the usual trees to hide behind, so I led him away from the road and started stamping the tall grass around a bush, hoping to show him that the desert was safe. A fat rattlesnake popped out, hissing in protest. We bolted back to the car. "Mom," he said. "I don't have to go anymore."

We arrived at the hotel. It had a small pool, and all of us were soon playing with the boys in the water. Vlada threw them so high into the air that it was hard to tell whether their laughter or their splashing made the louder noise.

Suddenly Max and Alex jumped out of the pool and pulled me aside, each speaking in an urgent whisper: "There's something wrong with Vlada."

"What?" I said. "What's wrong?"

"Why does he have hair coming out of his chest?" Max asked.

For all the comfort the three of us had found in our family of helpers at home, in the Widows and their children, I realized with a thud that Max and Alex needed more male figures in their lives.

I had asked a local club of amateur astronomers where the best place to test our camera might be. That night they invited us to their star-viewing party, a celebration of the new moon. We arrived at dusk at the old missile site. I looked up at the stars and felt my childlike wonder return. I think the boys felt it, too.

We set up the camera. We would have to wait until we were back at MIT to analyze our data, but our new type of detector, one not yet used for astronomy, seemed to do the trick. We knew at least that our experiment wasn't a total failure. In the company of my boys, my students, my camera, and the stars, I felt the flicker of an emotion in me that was so unfamiliar, I almost couldn't find the name for it. I felt *hope*.

The desert grew colder, and, beyond the concrete, snakes and scorpions were no doubt coming out. The amateur astronomy club left us. Max, Alex, and I stayed put, on the edge of the New Mexican desert and the Milky Way. We might have pictured fearful things, rattlers and worse. We didn't. The three of us stood out there, together on that moonless night, and none of us made a move to leave. We wanted to stay out there with the stars until the sun began its rise, washing them out one by one until even the brightest had disappeared.

We would know they were still up there. People talk about the sun and its reliability, how even on the darkest days we know it will come out again. A kind of opposite is also true. Even on the brightest days, beyond blue skies, there are countless stars shining over our heads.

•

At our next coffee, I told the Widows about New Mexico: how pretty the stars were, and how much closer we were to reaching them. I normally talked about ASTERIA with fellow astronomers or aerospace engineers, and they would ask me technical questions about the camera lens and projected or bits and the software that we would use. The Widows didn't care about those things.

"It sounds like you are traveling a lot with your kids," Micah said. She didn't say it meanly. She was making an observation, a point of order. I still thought I heard a hint of judgment in her voice.

"Yeah," I said. "It's getting expensive. I'm really blowing through my cash."

Almost in a chorus, the Widows sang away my financial concerns. "None of that matters," one said.

"Just do what makes you happy," someone else said.

Micah decided to offer a dissenting opinion. "Are you going to take them everywhere with you? That sounds exhausting."

I took a deep breath. I confessed my worst nightmare: that something would happen to me while I was away from them. Almost in the way that I imagine someone who is blind in one eye lives in constant fear of losing the other, I had visions that I would die and turn the boys into orphans. For some reason I was most worried that I would die in a plane crash. I knew how unlikely that was, but it seemed like a real possibility to me. (I had again met with my lawyer, Freya, and she'd helped me make an airtight estate plan.) I still struggled to leave Max and Alex at home when there was even the faintest possibility that it might mean leaving them forever.

"Oh, Sara," Micah said. "You have to move on."

There are moments when someone can say the simplest thing and it just hits you in a particular way. This was one of those times. I had heard those words often enough for them to become meaningless, but for some reason Micah's observation became a statement of fact. *The Earth is round. The sky is blue. I have to move on.* It was so basic and elemental—the meaning of life is movement, constant transformation—but it could not have sounded more profound to me. Her words found their way to a welcoming part of my brain. That part must not have been open when I'd heard them before.

Maybe that night under the stars had something to do with it. At the end of our trip, Mary, the boys, and I had arrived at the airport for the first leg of our flight home. Mary was carrying the ASTERIA camera in a hard-sided black Pelican case. The plane was surprisingly small, and when we walked out on the tarmac into a blinding sun, the baggage handlers insisted that Mary check the satellite into the cargo hold. She tried to explain that the case contained more than underwear and socks. It held more than a million dollars in parts and labor. I couldn't hear what she said over the sound of the engines, but her body language was out of character. By nature, Mary was calm and quick to smile. Now she was feverish, her eyes wide, her hands waving in the air. Our understanding of our place in the universe might be locked inside her battered plastic case.

The baggage handlers were unmoved: the case had to go in the hold. Mary refused. Being something of an expert in meltdowns, I went over to help the situation. Mary and I tried everything—we bartered, we cajoled, we begged, we demanded, we smiled, we admonished, we reasoned until we ran out of verbs to describe our shared efforts and time before the flight. I'm not sure how, but we finally claimed victory. The camera would be coming with us.

The only problem was that in the distraction my boys had vanished, and the plane was about to leave without either us or them. Mary and I looked around frantically until we raced onto the plane, the only place we hadn't searched. Max and Alex were sitting in their assigned seats, strapped in and ready for takeoff. I was relieved and a little embarrassed. I was also impressed.

"They're going to survive without you," Micah said, snapping me back to the present. "They're going to thrive without you. You have to get to where you can see that."

She was right. My boys were still young, only nine and seven. Puberty was years away. But they were well on their way to growing up. Just when I was really getting to know them, they were changing like my science, traveling at the speed of light. They were becoming capable. They were becoming responsible. *They're going to survive. They're going to thrive.* After all that Max and Alex had endured, my beautiful boys were still going to become good adults. They were going to forge the future.

•

One Friday morning we began confessing our shared Widow superpower: our ability to cataclysmically dissolve in public. I can't remember which failure to keep ourselves together was the most cathartic. I could have told the story about the library and the book I'd needed, or that time in the grocery store line, or when the Wheelers had taken offense at my leaves and my children's toys, or that visit to the hardware store, or that other visit to the hardware store, or that visit to the hardware store that came after that other visit to the hardware store. But I decided to tell one that involved William Bains, my imaginative British research partner.

William and I were invested so intensely in biosignature gases, we'd sometimes work on a Saturday or Sunday at my house, cramming as much as we could into each of his now quarterly visits from the United Kingdom. William had four adult children, and his experience with fatherhood helped him tolerate my boys, at least. One weekend, I asked him if he could spend some time with Alex. I wanted to take a break and play tennis with Max. One-on-one time was important. Attention of all kinds was important. We all went to the same park, but it was massive, and the tennis courts were at one end. The last I saw William and Alex, they were headed for a playground.

"Follow me," Alex said, and they disappeared.

When Max and I finished our game, we made our way back to the playground. William was flat on his back in the sand, his skin the color of milk. Alex was kneeling beside him.

"William! What's wrong?"

His response didn't quite land—I thought that he might have been joking, in the middle of playing a game. But then I realized he was in serious pain. He was fairly certain that he'd dislocated his shoulder. "Hang on, William!" I said, and I ran off to get the car. The four of us climbed in, William sweating oceans out of his forehead, and we raced across town to the local hospital. It was the same hospital where Mike had gone when he had first fallen sick. The hospital that had ignored his bad back and destroyed his ankle. The hospital from which he had been delivered home for hospice.

All those recollections surged back, carried to my frontal lobes by the smell of sickness and the beeping of unhelpful machines. The triage nurse asked William inane admissions questions. His eyes watered. He was in obvious pain, and all he needed was for a doctor to come out and put his shoulder back in. William, British and polite, did his best to answer her through his clenched teeth. I wanted so badly to grab her and scream in her face: *Just get him a doctor!* Then she pressed the wrong key on her computer and erased everything she had written.

"Oh dear, I'll have to start again," she said.

Her carelessness, her uncaring, and the ghost of Mike reflected in the floor tiles . . . Everything combined into the fuse of a bomb. I detonated spectacularly. William later said that he'd never seen anything like it. I ended up having to leave, apologizing to him and asking him to call me when he was done. I couldn't bear to be in that place with those people for a second longer. The next time he visited, I asked him to look after Alex again, "but be safe!" They settled on a staring contest.

I told the Widows about that day, laughing a little, crying a little more. Stories began pouring out of the rest of them. I laughed and cried some more. Someone had thrown down with another parent in a school hallway after an accidental violation of parking lot etiquette. A vacation for Melissa and her son in Puerto Rico had culminated in a collapse so complete that it sounded as though she might have made the news. (She'd had a breakdown at the hotel pool and, due to the language barrier, her explanation that her husband had died led to a brief murder investigation.) Chris told an agonizing story about losing her car keys in a grocery store—grocery stores are apparently the seventh circle of hell for widows—when she was already running late to pick up her kids. A store manager tried and failed to console her, and together they searched cart after cart. At last they found them, but not before, as Chris would have us believe, she had cried so hard for so long that other shoppers must have thought she'd lost her children, not her keys.

Before I became a widow, I had always imagined stoicism among the prematurely bereaved: thin-faced women wrapped in black shawls, standing on beaches, looking out at the sea that had swallowed their men without remorse. The periods of quiet sadness, I understood. I understood opening a drawer and crying into a husband's socks, or being bent over double by a picture in a yearbook. Our public freakouts, our constant charges against the ramparts of polite society, seemed less logical to me. We must have needed other people to know how much we were hurting. Or maybe we wanted the world to know that we weren't scared of being hurt anymore. Maybe we were like veterans saluting each other when they marched together in parades, through streets lined with the incompletely appreciative: *We know things that you never will, and don't you ever forget it.*

In June 2012, it was time for me to go to Tel Aviv to receive the Sackler Prize. I was still paralyzed with anxiety about leaving the boys, but I'd called their doting, witty aunt Rachel, Mike's sister in Alberta, and she had agreed to fly to Boston to look after them. At least they would be with family. Now a second dilemma arose, almost equal in my hurricane mind: I had nothing to wear. My work clothes consisted of casual outfits for the office and buttoned-up suits for everything else. I rarely had much call to wear a dress. I wanted to wear a dress.

I did what I had grown used to doing when the material world threatened to submerge me: I phoned Melissa. We talked through everything, which was now mostly my fear of leaving the boys, but also my fashion woes. The next day when I came downstairs, a garment bag had magically appeared in my front hall closet. It was stuffed to overflowing with beautiful dresses from Melissa's wardrobe. Going through that bag felt like shopping. I tried on dress after dress and imagined a different life for myself. I imagined a different me.

I chose a navy-blue one with short sleeves and three pearls sewn into its front. With Rachel taking care of my boys and Melissa taking care of my wardrobe, I felt as though I had been twice rescued. I flew to Israel and basked in the heat and attention a little bit. It wasn't an elaborate ceremony. I shared that year's prize with Dave Charbonneau; time had done its good work after all. We each gave a presentation, and I received a certificate that would go in my office. That was it. But I had put on Melissa's dress and known, if only for a single evening, how it felt to look at myself from the outside and see someone pretty.

I came home exhilarated, and also exhausted. I'd picked up some germs somewhere along the way and immediately fell sick. The boys demanded more than their usual attention after my time away. The house needed cleaning; the garden wanted to be readied for summer. It would be a long time before I wore heels again. I pulled my hiking boots back on.

Melissa had said that I could keep the blue dress. For months after, when I opened my closet and the light and my mood caught things right, that dress looked like a cape.

- - - - - - - - - - - - - -

CHAPTER 14

Sparks

Father's Day was a gorgeous day, bright with fluffy white clouds. It's weird, looking back, how often the sun shone on the Widows' gatherings. In the movies, there would have always been rain, but in reality, my memories of that time are filled with blue skies.

All of our children had become friends. They didn't gather because their fathers had died; they gathered because it was fun. There is a reason every children's book is written from the perspective of the child. Children don't care about adult concerns. We think of children as helpless when they are the embodiment of resilience, more impervious to outside forces than we could ever be again. Despite their suffering, our kids still knew pure joy.

Max went through a phase for about a year after Mike had died. I was never sure if he was offering an innocent observation of his world or trying to deliver a message to me. Maybe it was both. It usually surfaced on the drive to school. He'd say in one breath: "Families have one mom, one dad, one boy, one girl, one cat, one dog." He repeated it daily, the way I told myself that I was going to be happy. I explained to him that our family was a different family, but it was a great family, and we had a great life together. One day he just stopped saying it, and he never said it again.

On Father's Day we put out a huge lunch with a particularly impressive selection of sweets. "The Widows really like dessert!" Alex shouted. It always cracked me up when they called us "the Widows" rather than "the mothers" or something like that. There was a big tray of cupcakes with fortunes inside them. I can't remember which one of us opened the particular fortune—and thought it was a good idea to read it aloud—but someone read: *Treat each day as though it will be your last, because one day it will be.*

The words hung in the air for a moment. *Jeez, cupcake fortune. That's a little heavy.* But then we began smiling at each other. Someone started to giggle. Someone else started to laugh. Before we knew it, we had all dissolved into hysterics, following the trail blazed by our kids. Our collective sense of humor had verged on macabre from the beginning. Now it was getting really dark. We joked like coroners, like homicide detectives. Our lightest moments came in the face of death.

The kids ran off after lunch, and we all leaned back in our chairs in the sun. Our adult concerns resurfaced, like blood finding its way out of a torn-open scab. Every time we got together, it seemed our losses had become more tragic, not less. We had all become widows at different times, but we were all still in the endless middle of our grief. In strange ways, our husbands' deaths weren't the saddest parts of any of our stories anymore. (In my case, it was the little reverberations and ripples that came after, the triggering effects of Eastern Orthodox weddings or seeing a river rise in the spring, a canoe tied down to the top of someone else's car.) It wasn't the torrents that did it; our worst spells became less severe with time, and their arrival became more predictable. The greatest threat lay in the countless, constant trickles.

One of the Widows who had lost her husband to cancer had found the strength to go through more of his things. We agreed that the materials of a lost life could be hard to handle. It's incalculable, the heartache that can come from shoes at the door or a toothbrush by the sink or a hole in the backyard. Some of the Widows kept everything. Some threw everything away. I fell somewhere in between. I had filled that Dumpster, but a few of Mike's things were impossible for me to discard. His boats were the hardest, logistically and emotionally. He had given one away to a friend before he died, and I had tossed the broken ones. But then there were the Old Town Tripper, and my Dagger Rival, and the Royalite Swift Yukon that Mike had outfitted with four seats, a boat fit for a family. I didn't have the strength to look at them, let alone part with them.

This Widow had sat down at her husband's desk and begun sorting through the bottomless piles of papers, trying to answer the same question we all had to tackle over and over: What's still important, and what doesn't matter anymore? In between the tax documents and insurance claims, she had uncovered four plane tickets to Paris. It took her a minute or two to realize what she had found.

Her husband had bought them the day before he had started his chemotherapy, placing an optimistic bet against cancer. It had been a very private wager. He hadn't whispered anything about his plans to anybody. The plane had taken off with four empty seats the day after he had died. Those four unused plane tickets represented all the hope that was now lost, all the adventures that would never be had.

That's where we Widows found ourselves in our weakest moments: We were each trying to look forward, but too often we were given reason to believe that the best of our life was behind us.

The children were still off playing, and we could hear their distant laughter. They were just far enough away. They didn't see that Paris was in full flood.

•

A lot of widows move out of the house they once shared with their spouse; there are too many fingerprints, too many echoes in the halls. I had thought about it and decided to stay. I liked our pretty yellow house, and the boys had experienced enough change. I wanted them to have what I had only dreamed of as a child: one roof. The Widows were supportive of my choice, but they told me that I couldn't think of the house as Mike's and mine anymore. I couldn't think of it as *ours;* otherwise I would never be free of my sadness, as though I'd invited an emotional vampire through the front door. I had to make a clean psychic break. I had to make it *mine.* I had started making that transition almost accidentally, on the day I filled the Dumpster. I'd continued the process when I had Jessica's room painted lavender. But it was a big house. There was still so much work to do.

I had to make my bedroom my own. I decided to stay in the rainbow room across the hall from the boys. I wanted to be able to hear them if they woke up, and I didn't want to return to the bedroom that Mike and I had shared. I was scared of what dreams might come to me there.

Mike had continued appearing to me in my sleep; he was always returning from a long time away, another trip, another coma. He always arrived on the steps of the house, on the outside looking in. In the shock of his arrival, I struggled to collect my thoughts and find something to say, and before I could say much, he always disappeared. Sometimes the dreams were so vivid that after I woke up, I had to scan through my memories like files in a cabinet, remembering that I did watch him die.

I hired a decorator to take the hard but necessary decisions out of my hands. I found Bob. My tastes were becoming more feminine in the absence of Mike, and Bob seemed to understand what I needed. He stood in the room and announced that he wanted to see it painted a deep antique pink. I had a piece of pine furniture that I wanted to keep; he would find a four-poster bed to match. He would also find a loveseat, cream-colored and frilly, and he already knew which polished lamps he would bring in to complete the space. It would be indisputably mine, a palace fit for a princess. I smiled at him and nodded: *Let's do it.*

Not long after, the painters arrived with their drop cloths and rollers. I showed them into the room, empty now except for Max's yellow walls and Mike's rainbow. I stood and watched while the painters pried open the cans and poured the thick pink paint into their trays. Their rollers soon ran over the walls with quick, practiced strokes. It had taken Mike hours to paint his rainbow. Now it vanished in seconds.

The tears ran down my face in steady streams. I knew, intellectually, that the Widows were right. I needed to make forward progress. I couldn't spend the rest of my life drowning in grief. I had to kick my way back to shore. But when you lose someone, you don't lose them all at once, and their dying doesn't stop with their death. You lose them a thousand times in a thousand ways. You say a thousand goodbyes. You hold a thousand funerals.

•

The Widows told me that I wouldn't hold the last funeral until I had starting dating again. Near the one-year anniversary of Mike's death, Melissa came over to my house. She led me into the kitchen, made sure we were alone, and told me that I had to pretend, at least, to be interested in men again. Until I started dating, until I looked at a man with the intention of putting my mouth on his, my grieving would remain incomplete. I would always be looking behind me, taking stock of what was missing. I needed to see what else was out there.

I knew what was out there. Thousands of billions of planets, orbiting hundreds of billions of stars.

Melissa shook her head. Other men, she said. Lots of other men. "Now," she said, in a whisper that wasn't much louder than the hum of the refrigerator. "Even though you can't get pregnant anymore, you can still get STDs. You'll need to take precautions."

I just about fell over. I have no idea how old Melissa thought I was, but I was only forty. "Melissa, I can still get pregnant!" I said. I couldn't bring myself to say much more. I wasn't a teenager. I didn't need lessons in the dangers of unsafe sex. I didn't need to think about sex at all. How would I even find a man to go out with me? Who would want to date a widow prone to public displays of rage and crying fits in libraries and grocery stores? Just the idea of getting to know someone new, and letting them get to know me, was repulsive. I could barely look after myself and my boys. Melissa might as well have told me to find a unicorn, or an alien.

"You're not looking for perfection," Melissa said. "You're not looking for a husband. You're looking for some guy to take you out for dinner and then to bed."

I told Melissa how hard I had cried when the painters had

covered up Mike's rainbow. How was I going to react to a new set of lips?

"That's exactly why you need to start dating," she said. I had to reclaim my heart the way I was reclaiming my house.

"Don't worry so much," Melissa said. "It never works out with the first guy."

You're right about that, I thought. The first guy had died.

•

On good days I could see what I had, not what I had lost. Max and Alex were my principal vessels of optimism, and I continued to keep them close, like talismans to ward off evil, to shield me from my next wave of grief. I still had overwhelming anxieties about my travels, but not traveling wasn't really an option for me. I had to go to conferences and meetings. The stars won't come to us, and rockets don't launch from Massachusetts. Jessica or Diana could handle overnights, but anything longer asked too much of everyone, including me. The only solution remained taking Max and Alex with me.

In July 2012, I was speaking at two separate conferences in Europe, and I decided to use them to justify an epic three-week journey for us. The boys and I took along several of our usual coterie: Jessica and a rotating gaggle of students and postdocs, including Mary Knapp and Leslie Rogers, a former grad student who focused on the composition of mini-Neptunes. Sometimes I wondered whether I was making excuses to travel with so much company—before Mike's death, I didn't have any friends; after his death, I collected people the way a black hole swallows up any star that gets near it. Maybe it was some subconscious holdover from my childhood, from the ever-changing cast of babysitters that my father employed, or those nights when three of us slept together in a room at my mother's house. Maybe it was a natural aftereffect of the loss of someone who had once been so close. Whatever the reason, I liked being with people I could call friends. My students often presented their own work at the conferences, and I hoped that they got their share out of the experience. But mostly those trips were for me. I needed the world to feel smaller and its spaces less empty.

We started out in London. Jessica got terribly lost one afternoon because she didn't know how to pronounce "Leicester Square," where we were crashing in tired heaps at a colleague's apartment. Then we went to Paris for a couple of days. I stared at *The Thinker* by Rodin; the boys preferred playing with the pigeons outside the Louvre. We needed to make four different train transfers to reach Heidelberg, Germany, for one of the conferences. Given the size of our pack— there were seven of us just then—it felt like a more complicated exercise than a rocket launch.

In Switzerland, we were hosted by postdoc Brice; he was kind enough to bring us to the top of the Alps, near the windmills of Saint-Luc. The François-Xavier Bagnoud Observatory, striking with its shining silver dome, sits on the shoulder of a high mountain. The view of Earth from there is impressive enough: Rivers of green spill between rows of peaks like shark's teeth. At that altitude, the panorama is literally breathtaking.

Something called the Planet Path sits not far below. It's a scale model of the solar system (old enough to include poor, demoted Pluto), with a winding dirt trail between the planets. Every meter of trail accounts for one million kilometers of space. Even given that significant reduction in scale, the walk from the surface of the sun to the far side of Pluto is six kilometers long, or a little less than four miles. Leaping across hundreds of thousands of kilometers really gives you a sense of the size of space. The Planet Path encompasses only our immediate neighborhood, and it still takes stamina to cover.

Brice had once worked at the small observatory, sleeping each night inside its hopeful confines. There was a time when astronomers were like lighthouse keepers. They would climb up their mountains and look at the stars by themselves for months at a time. Now most small telescopes are automated, with robots programmed to find and examine certain stars on our behalf. But Brice had looked out from his mountaintop and watched stars like ships.

He had made an important discovery. Using the Transit Technique, he had seen passages of a planet that's known today as GJ 436 b. The signal he'd found was later confirmed by a larger Israeli telescope, but Brice was the first human to see a special place. GJ 436 b is about the size of Neptune—at the time, it was the smallest exoplanet yet discovered—and it follows an orbit that's almost impossibly close to its star. (GJ 436 b is fourteen times closer to its star than Mercury is to the sun, making its version of a year less than three days long.) It also orbits its star on a perpendicular path: a polar orbit. We now know of a few planets that climb over and fall under their stars rather than taking a more equatorial path, but GJ 436 b was something of a revelation. As though it weren't different enough, GJ 436 b also has what appears to be a comet-like tail, an exosphere that makes it look as though its atmosphere has sprung a leak. GJ 436 b is nothing like Earth, except that it's a planet, orbiting a star.

There we all were, vacationing in a stunning sculpture of a place, where my postdoc and friend had found a planet so different from ours that it almost defies illustration. The Romans

believed in a sky god named Jupiter. The Egyptians believed
that when their kings died, their souls became stars. Brice had
used an elegant assembly of glass and mirrors to see something no one had imagined. The lesson of his find is the lesson
of every discovery, and standing in those same mountains, I
made a promise to myself never to forget it. He had needed
only to be quiet, and still, and to keep his eyes open. He had
stayed in one place, but he had seen another. Through his window was a brand-new world.

•

The carefully curated list of Death Days was emailed out not
long after our first gathering at Gail's. We decided that we
would be together for all of them, too. Those anniversaries
were a way to mark our small victories and the passage of
time. It was also crucial for us not to be alone on them. They
could be a portal to suffering, to reliving our losses. They
could take us back as easily as they could help us move forward. We penciled seven more gatherings onto our calendars.

On July 23, we gathered to mark the date of Mike's death.
I couldn't believe it was a year since he had left. I couldn't
believe that it had been only a year.

It was a sweet summer evening. Melissa took us all outside,
where the kids squeezed together on the steps on my tiny
backyard porch, and she made a little speech. She was speaking to Max, Alex, and me, but she was also speaking to the
group. "You all know why we're here," she said. The kids
knew what was coming; they fidgeted in discomfort, pretending the problem was the heat. "We want to help Max and Alex
mark their first year without their dad. And although none of
your dads is here anymore, we hope—we know—that they
will still be guiding lights in all of your lives."

We had all started sniffling in the deepening dusk. Melissa wasn't finished.

"So we're all going to light sparklers!" she said.

She had brought forty sparklers for each child—an enormous number. (She had asked another Widow to bring some down from her vacation in Maine, and "some" had become "hundreds.") The kids were kids again. They raced up and down the driveway, lighting the wands two at a time. They had fistfuls of light, and whenever theirs went out, they lit up new sparklers off one another's still-burning flames. The air became thick with the unmistakable smell of sulfur, and a cloud of smoke filled the driveway and drifted down the street. Someone worried aloud that the fire department might show up. Someone else hoped they might; some beefy fireman eye candy would be a nice way to wrap up the evening.

That's how it always went, still. Up and down, backward and forward. There is nothing remotely linear about recovery. All of the Widows had setbacks, low moments when it felt as though we'd lost all the progress we had made. They could still come at the most unexpected times. One day I texted one of the Widows that I was actually feeling good for a change. "Just wait till the rain comes," she wrote back. "You'll feel terrible again." Sometimes they came at obvious times. Chris went on a date with a man who looked a little too much like her lost husband. She had to cough and sneeze and hide behind the menu to cover up her perpetually brimming eyes. She was still the only Widow who had joined after me, and I tried to pay forward the advice I'd received from the others. I told her that it was too soon for her to date; she'd know she was finally ready when she found men to be good-looking again. She'd reacted angrily—"I *am* ready, Sara," she said—but I knew the difference between wanting to feel something and actually feeling it. I had so many days when I woke up believing I had nearly made it, but there was a black spot still in me, and it took only the slightest knock or careless moment for it to become an awful, spreading stain.

I needed to find the boys a new school. Their Montessori school's enrollment was declining and its debt climbing, and it became painfully clear that it was about to go under. I followed Mike's directions in the Guide to Life on Earth, and arranged for an admissions interview at the school he had recommended. I hated that feeling of being in between stations, the uncertainty of one of those rare things that should be certain. You should know where your children go to school. I so badly needed things to work out.

Thankfully, the kids liked the look of the new school, and it was ready to welcome them with open arms. Our transition would be manageable after all. I had solved a major problem in our collective lives, and I had solved it mostly on my own. I had a physical response to the resolution. I stood up a little straighter. My eyes felt clearer. I took the boys back to their current school to finish that day's classes. It was sunny and warm that afternoon, and the world seemed to me to be bursting with green.

I don't usually like driving, but after I dropped off the boys, I really lost myself on the road. I rolled down my windows and turned up the radio. There was no traffic. My hair blew back, and for the first time in what seemed like years, I felt a genuine smile surface. I don't know how to explain that sensation, but my smile felt almost like a crack in my face, like muscles coming out of atrophy: a peel of thunder after a long drought, the first calm morning after a terrible storm.

I put my foot down. There is a road heading into Concord

that widens before crossing over the river. I imagined it was a runway, and I was preparing for launch. I lifted out of my seat.

That's when I saw the flashing lights of a police car.

I looked down at my speedometer. I was going 70.

In a 35 mph zone.

I pulled over. I turned off my engine and looked at myself in my rearview mirror. Every drop of good feeling drained from my body. My smile disappeared. My atrophy returned.

The officer walked up to my car and leaned into the frame of my still-open window. I made a quick decision not to fight the familiar hot rush of oncoming tears. Tears might not hurt me here. The Widows called it "playing the Widow Card." It was an accepted strategy.

"My husband died," I said, crying.

He took my license and went back to his car. I sat in my seat and waited, my black spot blooming like an oddly sinister flower.

The officer walked back up to my car. He had a piece of paper in his hands. He passed it to me along with my license.

"No ticket today," he said. He had just given me a written warning. "Please drive a little more carefully. Good luck with everything."

Now I was crying for a different reason.

Another night, I called Melissa, barely able to summon words. I was crying so hard that her son heard me through the phone. I could hear him in the background: "Did someone else die?"

Melissa dropped everything to come over, supplying me with another version of that beautiful dress. This time she brought me a picture book of shells. She had noticed I had a shell collection, gifts from my father and grandfather. It was her way of showing me that the world is still beautiful, if only we remember to look. She also gave me a large stone that had been smoothed by a river to the point where its rounded edges were almost soft. She told me to put it in my purse, which I did. Whenever I'd rummage around in it, I'd feel that stone between my fingers and be reminded that time can wear the edges off everything.

CHAPTER 15

Rocks in the Water

The last line of my Guide to Life on Earth was an explicit instruction: *Scatter my ashes on the Petawawa River.* A few weeks after the anniversary of Mike's death, after the sparklers in the driveway, I finally prepared to complete my last assignment. The mouth of the river was near his childhood home in Ottawa. The upper stretches were more remote, deep in Algonquin Provincial Park, and in the spring they became violent and churning. We had spent a week up there back in the summer of 1995, early in our relationship. I could still remember a fox I had seen dart through the trees. I knew I was ready to say goodbye to Mike when I began making plans to go back.

The Algonquin Radio Observatory rises near the Petawawa, built in a place of total silence. When Mike and I paddled the river, swollen and heavy with rapids, we had stopped there and found a space camp for children. The observatory had since closed. I learned that a young family had taken it over on a government lease and started fixing it up. When they arrived on the site, the front door was open and snow drifted in from the outside. Now they offered the telescope for rent.

They also had beds for overnight guests. I booked most of their available rooms. As was my new custom, I approached the woods with a traveling circus: Max, Alex, and me; Mike's mother; Pete, Mike's best friend and his companion on that last trip to the Galapagos; and Vlada, by then less a postdoc and more a trusted friend.

Before we left, I went to see Dave at the funeral home one last time. I told him that I was ready. He nodded and went to retrieve Mike. He returned with the most perfect box, made of wood that had been joined so well, it looked to be without seams. It was exactly what I wanted but had been unable to describe; I would never be able to thank him enough. Inside, he had separated the ashes into two plastic bags. There was a small one for Mike's mother and brother to scatter closer to the mouth of the Petawawa, where it meets the Ottawa. The larger bag was for me to take into the woods.

As one last service to me, Dave told a story. Human ashes, he said, aren't fine. Some are sharp-edged, with tiny fragments of bone that haven't yet been ground to dust. I needed to be careful when and how I scattered them. He said that he knew of a woman who had scattered her husband's ashes in their yard, and some of them blew over the fence and into her elderly neighbor's eye. Dave was laughing by the time he got to the part of the story where that man's eye had been infected by the remains of his dead neighbor. "Watch the wind," Dave said, almost pink with joy. I have no idea if his story was true, but it served its purpose. I didn't want Mike's final resting place to be inside me or anyone else. I would watch the wind.

We set out on the ten-hour drive north. I was not doing well. I radiated anxiety, and the children felt it and fed off it. They acted out from the beginning, complaining about the ride and each other. We pulled off the road for a snack, and

Vlada became the teacher, not the student. "Sara, you have to stop stressing," he said. "The kids are reading you. Stop it." I took a few minutes to make the mental switch to calm, or at least the projection of it, and it worked. The boys leveled off for the rest of the drive.

Mike's mother and Pete met us at the observatory. We were the only visitors that weekend. It was a quiet place. After a fitful sleep, I put Mike's ashes in my backpack, and Pete and I set out for the river. I didn't take the boys. I wasn't sure how I was going to react when I finally released Mike, and I didn't want to expose them to the depths of my grief. I left them in Vlada's care. It was sunny when Pete and I left, with huge white clouds casting giant moving shadows—another perfect day when the script called for rain.

But something was amiss. When we reached the river, the water wasn't running at its usual levels. One side of the river was impossibly low, low enough that Pete and I could walk down the riverbed. Islands had turned into peninsulas. We walked over the granite shelves where the water should have been, our feet now finding the same rocks that Mike and I had fought to avoid with the bottom of our boat all those years ago.

At last we reached what should have been a waterfall. It wasn't hard to imagine that when the water ran high again, the place we were standing would be lost under rapids and foam. Pete and I agreed that we had found the right spot. We looked at each other, and I lifted the plastic bag out of my backpack. There was Mike in my hands. I almost couldn't believe it. For once I couldn't understand the math: all of his spirit, his strength, and his energy reduced nearly into nothingness. I checked the wind, and Pete and I took turns tossing Mike's ashes into the river. We reached the last handful: the last time I would hold him. The clouds drifted over our heads. The trees rustled. The river found its way. I let go. I scattered the last of Mike into the water.

Pete left the next morning. Vlada, the boys, and I went for a hike. I didn't mean for us to return to that spot, but we did, as though the river had funneled us there and now back again. We sat together on the rocks. Vlada could see the tears welling in my eyes and knew what I was about to say next. "Max, Alex. Guess what this spot is?"

The boys knew, too.

We sat in the sun and the silence and our own private thoughts. I don't know what the boys were thinking. I wondered whether the water would ever be so low again; I wondered whether anyone would ever stand in the same spot, or whether this place would only ever be a river. I wondered whether Mike would be part of the rapids that we had paddled together, forever in a place where the rocks were worn smooth, like the stone in my purse that I held in my hands and knew had spent time in the water.

•

Melissa was on me about dating again. I approached Max and Alex about it, trying to be transparent without going into the messy details. Max responded a little harshly even in the abstract. "No getting married!" he said with a shout. "We can't leave the Widows club!" He had come to love our little gang.

I wondered whether Melissa might be right. Jessica's older sister, Veronica, had taken her place in the lavender bedroom— now painted light blue for her—so I had a couple of free evenings each week. I decided to be open to what the universe might deliver me. The father of one of Max's friends dropped him off one afternoon. He was a divorced dad and kind of cute; maybe it was time to practice flirting again. I asked him for help with a stuck drawer in a wooden table. I played the damsel in very minor distress. We spoke while he wrestled with it. I tried to remember how you make small talk. I asked him questions and pretended to care about his answers. Eventually he opened the drawer, which was useful. A little while later I had him over for dinner. Despite having had a lot more practice cooking, I was still far from being at home in my kitchen, and though I managed not to scare him away, we opted to go out for dinner the next time around. We went to Walden Pond and looked at the water. It was time to kiss a new man for the first time since I'd met Mike.

The pieces did not fit.

I kissed him and thought: *This must be what it's like to kiss your brother.* He was and is a nice guy, and a good dad, and he has a great smile. He was a kind and gentle reintroduction for me to the company of men. We ran into each other again later, when we both picked up our kids at camp. We smiled at each other and meant it. We just weren't a romantic match. I decided that if I was going to date, I was going to hold out for something I wanted more.

The Widows held a photo shoot. Melissa hosted. She told us to bring some nice clothes and makeup; Chris, the most fashionable member of our troop, loaned me a pair of beautiful shoes. Melissa had hired her famous friend Gigi—she'd photographed Chelsea Clinton's wedding—to take our pictures. Some of them were group shots, all of us together on the couch, laughing at Chris's dirty jokes. It was bawdy fun. A few of them were headshots, meant for our online dating profiles. Gigi was very good at her job. I had a lot of first dates.

I didn't have many second dates. Third dates were as rare

as comets. I told my boys that things weren't really working out. "Why not?" they would ask. Sometimes I was the problem. I was too awkward or too smart or too sad or too abrupt or too something. Sometimes it wasn't me. "Well," I said to Max and Alex when they asked about another failed date. "He wasn't quite smart enough and he was kind of out of shape."

The boys tried to be helpful, offering suggestions. "What about the man at the rock-climbing gym?" Alex asked.

"He's twenty-five years old."

"What about that guy you were dating before?"

"The dad from camp?"

"No, that other guy." Then he paused. "Oh yeah, I remember. He was fat and dumb."

•

That autumn my house needed a new roof. I understood the purpose of a roof. It keeps all of the things that fall out of the sky out of your rooms. I did not understand the *function* of a roof—how a roof actually completes the demands of its purpose. I had never given it a second's thought. The woman who owned the house before me, another single mother, didn't seem to have given much thought to it, either. Over thirty years of however many owners and storms, the roof had done its roof things and been ignored by everyone underneath it. But apparently shingles don't last forever.

Mike hadn't addressed roof repair in my Guide to Life on Earth, so I called Melissa. And she did what she usually did when I called her: She swooped in like an angel.

She called a bunch of roofers and, having narrowed it down to a chosen few, continued her interrogations via email. She finally sent me a long written report. (She knew better than anyone how my brain worked.) Together we chose someone. Then she took me on a walk around the neighborhood, and we looked at every roof along the way for inspiration. I was struck by the variety of roofs, the shapes and colors and patterns and materials. Melissa and I settled on beautiful little tiles, laid out in a pattern that I thought was suitably historic without being fussy. She called the roofer and let him know.

He seemed confused when he saw me with Melissa the day he arrived to start work. He had never spoken to me and had no clue who I was. He was also obviously taken with Melissa, gorgeous in a pair of tight jeans. He was carefully but seriously checking her out. That was one of the first times that I realized women our age could still be objects of sexual attention—that we weren't doomed to live in the shadows of younger women, begging for their leftovers. I was a bit grossed out by the man's gaze, but I also saw it as hopeful.

The roofer soon stopped sharing my optimism. Like a lot of strangers, he made up his mind that Melissa and I were partners, and that his odds of landing her, already long, now approached infinite. Alex, in one of his purer moments of observation and honesty, had once said: "Mom, you should be gay with Melissa."

Two of the Widows were successfully dating widowers. That made sense to me, and I joined a dating site for people who had lost their first husbands or wives. It might be the saddest place on the Internet. Eventually, I met a great guy: successful, smart, athletic, funny. I liked him a lot. We had a good time together. If something went wrong, he'd joke: "It's not like anyone's dying." But going out with me did something powerful to his own grief. It overflowed its banks, and it risked drowning me, too. One time I came home crying from one of our dates and my boys said, "If it's so upsetting, why are you doing this?" I didn't have a good answer. I asked Melissa how to break up with someone. She texted me long, detailed instructions.

Step by step, I broke up with the widower. After nearly a year of trying, dating began to feel like an unnecessary drama to me. My life was dramatic enough on its own. I told myself that I had enough love. My children and my growing circles of friends—the helpers, the students, the father figures, the Widows—gave me all the support I needed. I sat Max down. "You don't have to worry anymore," I said. "I'm going to be in the Widows club forever."

On top of the Green Building, where I work at MIT, a satellite dish points at the sky. It looks impressive, but it hasn't been used much since the 1980s, its former function as a Doppler weather radar now obsolete. The MIT Radio Society, a 100-year-old club, occasionally bounced a signal off the moon with it. The rest of the time, it sat silent. That bothered me. Here was this instrument of connection, a means to turn satellites into allies, and it was incapable of collecting anything more than rust. I wanted to bring that unmoving dish back to life. I had the idea that I could use it to send commanding signals to my ASTERIA prototype once it was in orbit, and receive data from it in return. Later, I might even be able to direct the entire ASTERIA constellation from my office.

I went to go see the dean of the School of Science, a physicist named Marc Kastner, to ask if we could find the funds to fix the dish. We met in his office in one of the older buildings on campus, its warm wood paneling polished in stark contrast to MIT's cluttered labs. I knew Marc, but we weren't especially close, and we were both more comfortable with our work than with people. That's why I was so surprised when he asked me, the way Melissa had during our first fateful phone call, the easiest, hardest question: "How are you doing, Sara?"

I couldn't tell how much he wanted to know. But Marc has a friendly face, with a bushy mustache and genuine smile. I decided that I could be honest with him. I forgot all about the satellite dish on the roof and told him that I was still struggling under the weight of everything. I loved MIT, but the fall semester was demanding. Summer had meant that I could breathe a little; I had room to maneuver. If something didn't go quite right, I had time to recover. Idle minutes could be used to find remedies and make corrections.

Now I didn't have any margin for error. Every day felt like the subject of a mad science experiment: *How much can I squeeze into twenty minutes? Can I get out of my bed and back to it without falling apart somewhere in between?* I had mastered some of Mike's former responsibilities; others still eluded me. In between the jobs that I could do and the jobs that I paid others to do were the jobs that I half did. Even with help, every day was a kind of mountain. By the time I'd made the boys breakfast and lunch, driven them to school, put myself together for work, and taken the train into the city, I was already drained. I just couldn't keep up. "Something has to give, Marc," I said. "It might have to be my work."

I meant it when I said it. That wasn't the first time I'd thought out loud about quitting. It was a fairly constant refrain in my head. This contemplation had a different quality than my ennui at Harvard, when I had thought about becoming a veterinarian instead. I loved my work. I felt as though I was getting closer to achieving something that would bend forever how we regarded ourselves. But Mike's death had made me realize the mistakes I'd made in allocating my time. I was still mindful of the promise to him that I had broken; I felt a flash of shame whenever it came to mind. I had vowed not to make the same mistake with Max and Alex. They needed me more than I needed to be a few sentences in the long history of the world.

A few months earlier, four of the Widows had come to see me at work for one of our Friday-morning coffees: Gail, Micah, Melissa, and Chris. Only Chris swung by my office with any regularity. She took business classes at a college in the city, and on Wednesday afternoons she had a break in her day. She'd decided never to return to work as a data analyst, determined instead to strike out on her own, begin a new career perhaps in fashion or design after she graduated. I was so proud of her, and I always told her so. She was dabbling as a personal shopper at the time, and I became one of her first customers, a cheerleader as much as a client. She showed up with armloads of clothes for me to try, and I marveled at her ability to find the right things for me to wear—clothes I never would have dreamed of buying on my own.

Now I was the one who needed help with my career, and I turned to my four friends. I felt as low as the winter sun. Boston looked icebound from my window, the Charles River like frosted glass. I had my back to my blackboard. It was covered, as usual, with equations and diagrams, the mess I made and still make whenever I'm trying to translate my imagination into something real. The Widows were sitting across from me, almost like a hiring committee. Not one of them looked over my shoulders at the blackboard, the way visitors almost always did. The Widows never wandered. Their eyes looked only at mine.

"I think I have to quit," I said. "I can't do it. I can't keep up."

I expected them to protest. I expected to hear them admonish me, to tell me how lucky I was to have my job. It paid well. I was successful. *We were all dealt the same bad hand, Sara. Get a grip.*

The Widows didn't say anything like that. Almost in unison, they said: "We believe in you, Sara. Do what's right for you and your boys. Whatever you choose to do, you'll succeed. You'll be great no matter what."

I was stunned. Here were the smartest, strongest, funniest women I knew. I looked at them and I saw THE AMAZING WIDOWS OF CONCORD, superheroes in the never-ending fight against sadness, defenders of joy against the diabolical forces of grief. Part of me had still thought they must see me as a freak, even after all our time together; I still felt apart from them, though other than my boys they were the closest I had to family on Earth. I didn't realize until that moment that what I saw in them, they might also see in me.

Now Marc Kastner sat in front of me. He didn't say that I would be great at whatever I did. Leaving, he said, would mean that I'd be giving up the one thing I was most meant to do. He waved away any idea that I might abandon my search, and took a more practical approach. He told me that he and his wife had both always worked, and they couldn't have done it without a housekeeper. "You need a housekeeper," he said.

I told him that I had people—Jessica, Veronica, Diana, Christine—but for all the support they gave me, there were still jobs only I could do, and I couldn't afford to pay them any more than I was already, or for much longer. My savings had been drained. Marc nodded. My problem—again, like all problems, like every problem—was a problem of statistics. I had help, but I needed more. That would mean I'd need more money. He put his hands on his lap in front of him. "How much money do you need?"

Another question I had trouble answering. Sometimes complicated questions have easy answers: *Can we fix the satellite dish?* That's a yes-or-no question. In my world, simple questions were the hardest questions of all. *How did you sleep last night? What are you going to eat today?*

"Sara?" Marc said. "How much do you need?"

Marc gave me enough. I'm not sure how he managed it, but he found the money for me to pay for more help. From then on, there would almost always be someone in my house—the company that I was paying to keep, and that in return kept me. If nothing else, I would have a little more time each week to breathe. His generosity also told me that even at MIT, at a factory of impossible dreams, my dreams were worth fighting for. On a campus where people were trying to cure cancer and give machines emotions and make spray cans of artificial skin, the woman looking for aliens was someone worth saving.

Not long after, in December, *Time* magazine named me one of the twenty-five most influential people in space. This felt a little different from a physics prize that mattered only to other academics. The search for other life was becoming more legitimate. In a sliver of a century, we'd gone from members of my own community not believing that we'd ever see an exoplanet to the pages of *Time*.

I liked the picture they chose. I'm wearing black, of course, but I have red lipstick on. The photographer wanted a background that said: MATH NERD. I'm standing in front of my office blackboard, filled with my frenzied algebra. The caption gave me the title of *Earth-Twin Seeker,* which was cool, although Elon Musk got *Rocket Man* and Michael Brown got

Pluto Slayer. The last line of my biography didn't leave much doubt about my beliefs: I wanted to find other life to "bring the Copernican Revolution full circle: Not only is Earth not the center of the universe; there are lots of other living planets out there as well."

When I arrived for that Friday's coffee with the Widows, there were cupcakes with candles in them and *Congrats!* written in icing. I had to be told that we were celebrating me.

The conversation soon returned to life at home. It was still hard for science to compete with the concerns of balancing checkbooks, with the perils of pursuing younger men, with another call from another teacher, concerned about the morbidity of a piece of children's art: a cemetery filled with goblins, a sky painted black. In some ways, space would always matter less to me than it once did. What mattered that day was how well I'd done the laundry, or that the laundry had been done. I found pride, and maybe a renewed sense of determination, in each bill that got paid on something like time, each piece of chicken that came out of the oven edible.

But then I looked at those cupcakes. I didn't always feel capable or optimistic, but I had been given another reminder of the value of my *extra*terrestrial work. *Keep going,* I thought.

CHAPTER 16

Starshade

I woke up crying on New Year's Day. I had endured my first calendar year, 2012, without Mike. I remembered that terrible night, now two years gone, when we sat together at the kitchen table and took stock of our losses, soon unbearably to mount: "Next year will be worse." I fought back with different memories, remembering my father and his unshakable belief in belief. It had become a ritual for me by then, the practice of positive thinking. *I can be happy one day,* I whispered, a decibel louder each time. *I can even be happier than I've ever been. I might know a greater happiness than I did before.*

On January 4, with the boys headed back to school, and snow in the trees, the Astrophysics Division at NASA put out a call for applicants to join two new Science and Technology Definition Teams, or STDTs. (The fine people at NASA are good at space stuff, but they are true world leaders in initialisms and acronyms.) STDTs are committees of scientists, engineers, and academics—chosen from the ranks of experts within and outside of NASA—formed to tackle a specific, challenging project. I had first joined something like one back in 1999 at Princeton, to work on the ill-fated Terrestrial Planet

Finder initiative. The committee dissolves after its job is done—or, as I had learned, left undone, because space is expensive. Space also demands perfection and is a very hard place to be perfect.

NASA was putting together two teams for a pair of space missions they called "Probe-class." By the otherworldly standards of space exploration, these would be modest projects. The goal was a workable, buildable piece of hardware targeted at a billion dollars; the far-reaching Terrestrial Planet Finder had been many times costlier, by a factor of perhaps five or ten. A billion dollars was still a lot of money, but in the scheme of governmental budgets, it was measured. A new aircraft carrier costs about $13 billion. For a relative fraction of that, a good team working together on a worthy Probe-class study could make a miracle.

The teams were going to examine two sides of the same problem: the ancient dilemma of trying to see a dim light next to a far brighter one. The first team would explore building a space telescope with an exquisite coronagraph of unprecedented sensitivity—an internal means to block out a star's brighter light, the way we had tried to approach the problem all those years ago. The second would work on an external solution: the giant shield that had caught my imagination while I worked on the Terrestrial Planet Finder. It was time to try again to bring the best in science and engineering together. I still loved ASTERIA, but it was a backup plan compared to the sort of magnificent machines we might build. My old, shelved dreams felt in range again.

I decided to work on the shield, because it represented something monumental to me. The human eye, as brilliantly designed as it is, doesn't have a perfect internal coronagraph. Our pupils dilate, but they still have to be open for us to see;

darkness only comes when we close them all the way, trading one kind of blindness for another. We've augmented evolution's gifts by inventing hats with bills and brims. We wear sunglasses or draw the curtains. Why couldn't we make something that would serve the same function for a space telescope?

I reread the proposal, and I found my confidence rising. I examined this self-belief. It wasn't irrational. It wasn't hubris or the product of a desperate hope: *I need this*. It was fact-based. I tried to imagine the sort of person who would be useful on such a committee. It was me. What sort of brain? It was mine. I knew darkness, and sometimes you need darkness to see.

But reading through the proposal one more time, I felt a budding anger, too. What bothered me was a small note about how the committee would operate, the sort of thing that most people would read right past but that stopped me: The team, it was decreed, would meet face-to-face every quarter.

I stared at my computer and fumed. The expected duration of the project was eighteen months. That would make for six separate meetings, mostly in California. Six meetings, far away from home. I could manage, if barely, the existing travel demands of my work. Leaving home six more times, for multiple nights each time, would guarantee the collapse of the delicate balance that I was forever trying to find. I would always have to be bad at something.

If the people who had written that proposal had walked through my door at that moment, they would have witnessed the hottest of meltdowns. I would have scolded them for the cold, universal presumptions they made about how the rest of us might live and work. My children didn't have two parents or any other extended family nearby. They had me, and I had them. I had always loved them; now I liked them, too. I liked watching Max work his way through a math problem, or build something beautiful and intricate with his LEGO, or survey some strange given about the world and regard it with his wry smile. I liked playing tennis with him and feeling the warmth of his serenity on our walks home afterward. I liked watching Alex perform for a group of adults, his fearlessness when it came to strangers and heights. I liked hearing big words come out of his mouth and wondering where and how he had learned them.

I didn't want to leave my boys anymore. I had compromised enough. I told the organizers of the Probe-class studies that their call for applicants, however accidentally, was discriminatory. One of them—he didn't have kids—replied that he hadn't thought twice about the travel requirement. Of course the committees would meet. That's what committees do. To his credit, though, he saw the problem, and he saw the value in people like me, people who might be able to contribute but would not apply. He asked me how future calls might be more inclusive.

My answer wouldn't be mistaken for brilliance: *Don't confuse scientists with astronauts.* Wanting to see and wanting to go are two different desires, and not everyone is willing to sacrifice so much for one concrete achievement.

I was still upset, and I needed to vent some more. I had long stopped hearing the internal whisper that reminds us to be polite. I had been asked by the *Huffington Post* to write regularly about women and science; I hadn't yet taken up the offer. Now I started typing. I wrote about my predicament, which struck me as a permanent, irreparable condition; I had the terrible feeling that I would always be trapped between my loves. My post was published on January 14, 2013, under the headline: SO MANY EXOPLANETS . . . SO FEW WOMEN SCIENTISTS.

I began by writing about the explosion in our understanding of the universe, the celestial maps we had begun to draw. (Start with the good news.) Thanks to Kepler, we had determined that one in six stars like the sun hosts a planet the size of Earth. We had decided that seventeen billion Earth-size planets orbited their own suns in the Milky Way alone. Think about that. Seventeen *billion*. But most of them had been found by men. Why was only half our species doing nearly all of the job? Exploring space is a titanic quest. If we ever want to achieve what we believe we can, we will need every possible set of eyes. All of us deserve the same opportunity to help.

I believe that hundreds to a thousand years from now people will find a way to travel to the planets orbiting the nearest stars and will look back at us as the generation of people who first found the Earth-like worlds, I wrote in a torrent. *And, hopefully, long before that distant time we will have achieved equality for all humans.* I felt like standing on a street corner and shouting at the top of my lungs: "There are so many different ways to see!"

Despite my vocal misgivings, I still applied to join the committee. I had enough standing in my field that I felt certain I'd receive an invitation, but the choice whether to accept would be mine. I'd have to decide once again between my family and my work, between my desire to find another Earth and my wish to live my life to the fullest on this one.

I tried to fall asleep that night and instead stared a hole into my ceiling. Through it, I saw the gleam of a giant, beautiful shield, hanging in space as though from an invisible string, and connected to a shining silver telescope, two stunning spacecraft working in perfect concert to erase star after star, *there goes the Big Dipper, and there goes Cassiopeia,* and in their place thousands of new constellations, the invisible collections that exist today only in the palaces of our imaginations, in those private museums we have dedicated to the safekeeping of our smallest lights.

●

An astronomer named Lyman Spitzer is credited with the idea of building a shield to work in tandem with a telescope. His name might sound familiar—the Spitzer Space Telescope is named after him, because he was also the first to imagine the space telescope. That was in 1942. It took Spitzer twenty more years to realize, in a paper published in 1962, that a space telescope could be paired with what he called an "occulting disk." The thinking was simple enough. If we paired a telescope with a protective partner—if they could somehow be tied together but orbiting tens of thousands of kilometers apart— the shield could slip into place between the telescope and a star. Smaller, neighboring lights would then twinkle into view. In that same paper, Spitzer proposed the telescope that ultimately became Hubble. We built the telescope. We still haven't built the disk.

There are several reasons why. Humans have been building telescopes for hundreds of years, and space telescopes for decades. An external coronagraph is a brand-new machine. Astronomers hadn't decided on the best shape for it, and engineers hadn't invented the right materials with which to build it. Flying two spacecraft in unison is also hugely difficult. A telescope and its shield would need to be perfectly aligned for everything to work properly, and perfect alignment isn't something that "just happens" in weightlessness. How do you fix two floating objects in space? And how do you make them so that you can move them to look at another star and fix them again? And again and again and again?

After Spitzer's paper was published, progress came in its customary fits and starts. A burst of activity and attention yielded some small but essential gain. And then . . . Nothing, sometimes for years. Looking back, it was as though our community was working together on a million-piece jigsaw puzzle. Someone would come along and put a few pieces into place, and then someone else would take a look and fit a few pieces more. There were still so many pieces to go. Complicating things further: Normally when you work on a puzzle, you start with the edges, tying together the four corners. With the shield, progress seemed to come from the middle out. We didn't know if the final puzzle was a circle or a square or some other shape. The edges were what we were missing.

I remembered a decade ago, during our Terrestrial Planet Finder days, when we were first introduced to the star shield. Not long afterward, we heard a second, soul-stirring presentation by Charley Noecker. A small team of astronomers and engineers, mostly from Northrop Grumman, had been hard at work, and he wanted to give us an update. A paper espousing a version named BOSS (Big Occulting Steerable Satellite) had been published, advocating a huge, square-shaped screen, opaque at its center but transparent at its edges. Some scientists had argued that it would be impossible to get that shading exactly right, and Lyot's diffraction rings would remain a problem. Noecker and his colleagues agreed. They were among those eschewing the square.

He unveiled a flower-shaped shield to us instead. From the start, Spitzer had wondered whether something like a flower might prove the best shape for his disk. Not only would the petals help eliminate the ripples of light that bend around a circle or a square; they might diffract light in patterns that would yield an even darker dark. Now Spitzer's insight had been confirmed. As with our coronagraph shaped like a cat's eye, the inspiration for the new star-shield shape had been born of mathematics but was also found in nature.

A flower.

Years later, I could still close my eyes and hear that room during Noecker's presentation: It was cast in a pure and total silence. Our meetings are usually far from quiet. There are whispers of dissent that sometimes expand into loud arguments; during boring sessions, papers begin to rustle or laptop keyboards are tapped. There is coughing. But on that afternoon, there had been a vacuum. A void into which sound disappeared, replaced by a luminous agreement. *A flower. Of course. It should be shaped like a flower. It was always supposed to be a flower.*

But what shape the petals? And how many? That's what I remembered most of all. I remembered sitting in that room in the silence and thinking: *There are so many kinds of flowers.*

•

The first time I worked on the shield was also the first time I found myself torn between my roles as a scientist and as a mother. It was 2003, when I was at the Carnegie Institution. I had gone to a Terrestrial Planet Finder meeting when I was extremely pregnant with Max—probably a little too pregnant to be focused on trying to find other life in the universe. I accepted the congratulations of my colleagues, but it felt strange to be congratulated for doing something that women are so expressly designed to do. Nothing had ever come to me so naturally.

Max was a very new baby when I got a note from Webster Cash, an astronomer at the University of Colorado. (There is nowhere nearer on Earth to the stars than the mountains.)

After BOSS, and after a similar proposal named UMBRAS had come and gone, Cash was next to pick up the baton. He didn't want to make a pure shield, a star blocker. Instead he envisioned what was essentially a pinhole camera of almost unbelievable dimensions, an opaque square as wide as a football field is long, with a ten-meter-wide pinhole punched through its center. He asked me to join his efforts. Cash invited me along because he needed someone to make a model of the atmosphere of a simulated Earth. My coding abilities and my then-burgeoning knowledge of biosignature gases made me a good choice. I was, perhaps, Cash's only choice. There weren't many people doing what I did. I didn't mention to Cash that it might not have been the best time for me to embark on such a big project.

Luckily, Max was a good sleeper, and Mike had already started taking care of more than his share of the rest of our world. I worked mostly when Max was napping, when I could get some blocks of quiet. Once I stayed up all night to meet one of Cash's deadlines, working on atmospheric simulations, stopping only periodically to breastfeed my stirring baby boy. I'd rock in a chair with him, nurse him back to sleep, and then return to my code, to my algorithms, to my second, third, and fourth Earths. I was exhausted, but I was also content. I was helping to make two great things.

We made a successful proposal to study the possibilities for launching a giant pinhole camera into orbit. Jon Arenberg from Northrop Grumman reached out to Cash to see whether his company might help develop the screen; they were experts in building what we call "large deployables" in space. Their meeting came to a surprising conclusion. Jon convinced Cash to scrap his pinhole camera and design a smaller shield, returning to something much closer to the original vision but with a precisely calculated shape. Cash understood, as most mathematicians did, that the ideal geometry for a shield would be the product of an equation, a simple, elegant, provable sum. He also feared it would be nearly impossible to solve.

He spent the next several months working on the age-old problem of seeing the firefly next to the searchlight. Cash is the classic astronomer, hooked on space from an early age. He was eight when his principal obsession switched from dinosaurs to the stars. He has never looked back, only up. Today he has silver hair and a silver beard; he keeps pens in his shirt pockets. His résumé is impressive: He helped design parts of Hubble and other major missions. The star shield was his next grand adventure. He first tried dozens and then a hundred different designs, mostly flower-shaped, writing code and running elaborate, fruitless calculations. He never managed to block enough of the star. There was always interference. There were always rings of light radiating beyond the edges of his shield.

Cash had succumbed to the scientist's version of tunnel vision. He was stuck on a particular type of flower, something like a lily. Each of his designs had petals that radiated from the same small central focal point. In 2005, nearly two years after he had started work, he asked himself a new question: Why not a sunflower?

He started fresh, with a design that included a large disk in the middle, with the petals attached to it rather than to each other. He experimented with designs using twelve or sixteen petals, ballooning halfway along their lengths before tapering to fine points. The shapes of those petals changed in subtle ways, but those small changes mattered less than his choice of flower, which seemed to work. His calculations proved that a sunflower, thoughtfully shaped, would block out exactly enough starlight to expose any Earth-like shimmer next to it. He proposed a final design that would be about fifty meters across, made of light, durable materials, suitable for a life in space. It was, by most reckonings, the most practical shield design ever devised. Cash made a model of it, delicate and precise. When he held it, his eyes betrayed a flash of almost paternal pride.

Cash invented a new word to describe his version of the shield: "Starshade." He led a serious proposal to NASA in 2006, suggesting that the Starshade be built and launched alongside the James Webb Space Telescope, then scheduled to rocket into the black in 2018. They would orbit fifty thousand kilometers apart, kept in constant contact via radio waves. Cash thought we might see another Earth with them. We might divine distant seas. Clouds. Water. Life.

NASA listened. Then they rejected his proposal. Cash returned with proposal after proposal, but each was rejected. He couldn't begin to understand why. First the Terrestrial Planet Finder was canceled, now the Starshade was cold-shouldered. He decided that NASA was too risk averse. They knew they could build space telescopes; they didn't know if they could successfully build the Starshade. Around the country, small pockets of researchers at Northrop Grumman, Princeton, and the Jet Propulsion Laboratory in California fought hard to invest in the technology, to keep the flame lit over the years to come. But the Starshade never came anything close to being built. NASA never did say yes.

Instead, years later, in January 2013—boys back to school, snow in the trees—they put out their call for applicants for two new STDTs, one seeking to blot out the stars from within the telescope, the other from without. I applied to help build the shield, burned a little hot while doing so, and waited for fate-turning word from NASA.

•

The news arrived that April in an email from Doug Hudgins, a physical chemist at NASA. Doug is the sort of person who builds his own telescopes. The last sentence of his official bio reads: "His telescopes include a 24-inch f/5 Newtonian (home built) and a Meade seven-inch f/15 Maksutov Cassegrain." He is a fellow obsessive, a true believer in the search for other life in the universe. I was in my office at MIT when word from him landed in my in-box. It started with *Dear Sara*. That seemed promising. I kept reading, and my mouth began to drop open.

> I am sending this message to let you know that we have completed our review of the applications submitted in response to our call for membership on the two Exo-planet Science and Technology Definition Teams (STDTs) that we are establishing. In that regard, I am pleased to inform you that not only would we like you to serve as a member of the Starshade STDT (Exo-S), but we would like to invite you to Chair that team.

A few lines later, Doug concluded with an almost disarming humility:

> I would appreciate it if you could confirm your willing-ness to serve on the Exo-S STDT, and also let me know if you are willing to accept our invitation to Chair the Exo-S team.

I pushed back in my seat and breathed. *Willing to accept our invitation to Chair.* Seven words had rarely meant so much to me. Leading an STDT is a massive honor. The timing of this particular STDT meant that it was also a big responsibility: The Starshade might finally get built because of our efforts. Or it might never get built—because of our failures. We might be the fatal last flicker of Lyman Spitzer's dream.

I saw a lot of qualities in myself, good and bad, but I had never seen myself as a leader of something as monumental as the Starshade, especially not at such a pivotal moment. I knew I was smart enough. I had put in the work, and I had all the necessary late-night visions. But in my mind a leader was confident, poised, organized, in control. I was still uncertain, prone to meltdowns, flustered, and barely keeping things together. I never did have a capacity for self-delusion. I was brutally honest, with myself most of all.

I had been recently reminded just how close to the edge I now lived. I had taken the boys on a working vacation to the Pacific Northwest. Flying home through Toronto, we had to clear American customs at the airport. The immigration agent was unusually chatty, asking Max and Alex about their adventures. Then he turned to me: "Does their father know where they've been?"

"He died," I said, doing my best to maintain my composure. In an instant we were under dark skies. "Would you like to see the death certificate?" I asked. That's one of the secret burdens of widows: They have to carry proof of their suffering for the authorities. *Passport, plane tickets, death certificate.* The agent shook his head, obviously rattled.

I couldn't fool myself into thinking that those memories were buried. They were always right there, waiting, and now I worried that they might surface in some terrible way, causing me to lose the focus that the Starshade would demand. It was impossible for me to know what might open the gates or how wide they might be opened. I'd written in my *Huffington Post* op-ed that women in science—as in so many fields—often have to work harder than their male counterparts, especially to earn and keep positions of leadership; they bear higher burdens of proof. They are suspected as too emotional for such cool, clinical work. Too fragile. Too shrill. Never mind that men in my field were and are capable of outbursts and tirades. Of course they are: They are passionate human beings, invested in something they love. But I wouldn't receive so understanding a judgment. I was always careful not to show my emotions at work; working on the Starshade, I would have to be even more careful.

I was still sitting at my desk when I started building the counterargument to my concerns. I made the case for me. I had, I reminded myself, expected an invite to this group, albeit as a rank-and-file member. I remembered my formal education, all those years in classrooms and labs, researching, coding, building. I had made major advances in the field of exoplanets. But I had also gone on a different journey, covering a different kind of distance. I had lost my husband, and I had survived. I had built a new life for myself. I had made a new kind of family. I had found new friends and new places. Not many people had as much experience as I did with *new*. I could help make a new piece of hardware.

I said yes.

They passed along the names of the rest of the team. My phone rang almost immediately with a call from one of them: Webster Cash. He said that there must have been some mistake. Obviously he should be the chair. He was older, more experienced. He had been working on the Starshade for years.

He knew more about it than I did. He knew the right shape. He had even invented the word for it: *Starshade*.

It was a gut-check moment. I had always liked Cash; I still do. He's an insightful, gifted astronomer. I also understood the pain of losing your grip on something you've worked hard on, something that feels like *yours*. It had happened to me. I might have invented transit transmission spectra, for instance, but dozens of astronomers have come along since and done more with it than I could. You can't copyright an idea—all you can do is remember that sweet time when it was yours and yours alone. Then you have to let it go. You're not the only one who's allowed to look at rainbows.

"You have to call NASA," Cash said. I had to tell them that I didn't want to be chair and wanted him to lead the study.

"But I don't. I want to lead it."

Cash lost his temper. He raised his voice at me over the phone. He couldn't see that he already had his answer: NASA didn't want him to lead the study. NASA knew what he could, and would, give them. They had seen his proposal, more than once, and they had rejected it each time. He couldn't understand why, but they had to have their reasons. They wanted a fresh perspective, someone who knew what she didn't know. They wanted me.

"They want you because they know they can manipulate you," Cash said. "They know you're a pansy." Click.

A pansy, I thought. *What a pretty flower.*

•

Our first face-to-face meeting was scheduled for early July. It would be a two-day joint meeting of the two STDTs—the builders of the coronagraph and the shield. We would share the same foundations before we followed our separate pathways toward our collective dream of finding another Earth. Our days would be scheduled to the minute with lectures and brainstorming sessions. In addition to helping run the joint meeting, I would be given a half hour before lunch on the second day to present what I knew about the spectroscopy of super-Earths: thirty minutes to talk about my lifetime studying tiny, distant lights. I had a little less than three months to prepare to lead the Starshade team. I would do what I always did. I would read, and listen, and do the work.

I decided the Starshade would be a new beginning for me. It was everything I loved in a machine—an extension of our best selves, an instrument, like a paddle, of forward progress, of our most human desires. More important, I felt it was our single best chance to take clear pictures of distant worlds. It might provide us with proof of other life that we could see, knowledge that we could hold in our hands as well as our hearts: *We're not alone.*

The Starshade could also slip into place between my past and my future. I had spent too long thinking about before; it was time for me to focus on the after. I figured that if our committee did our job as well as we could, we might be able to use the Starshade to explore as many as one hundred star systems in my lifetime—statistically, enough to see perhaps a dozen possible Earths.

First, however, I had to steep myself in the Starshade's byzantine, sometimes unkind history. Unfortunately, there are lots of reasons for complicated things not to work. Our critics have hundreds of years of error and tragedy on their side. Most of our greatest successes started with explosions. When you're trying to do something that's never been done, there are endless arguments for why it never will be. I can construct the mathematical case for other life, that there are too many stars for us to be alone, but the argument against this is still simpler, plainer: Then why *are* we?

I told colleagues and friends about the Starshade. It was widely seen as that special kind of impossible—actually, irrevocably not possible. I heard the same message over and over again. *Give up. Don't bother. That's a doomed idea, long past dead.* I was at a barbecue for exoplanet scientists in the Boston area, and I was told repeatedly, between mouthfuls of hot dogs and hamburgers, why the Starshade could never work. (Scientists and social graces don't often exist in the same environment, especially not with food readily available.) *Two spacecraft flying in perfect formation tens of thousands of kilometers apart? A giant shield crafted to a precision measured in microns? Making stars disappear? Think about it, Sara.* To them I had been asked to climb a mountain without a summit.

There were rumors even within our team that we had been built to fail. I didn't like to believe something that cynical, but a small part of me saw the logic. If we couldn't come up with a workable solution, on time and on budget, then NASA would have the cover to shelve the Starshade forever: *We put our best minds on it, and they couldn't crack the code.* It would be like voting for a political candidate just to prove how awful their proposals might be. *We tried that, and it didn't work. The other answer must be the right answer.* The other answer would be the coronagraph, unless they, too, failed. Then our only end would be blindness.

•

For some reason, Max and Alex had decided that the front hall was the perfect place for them to wage lightsaber battles.

(If two kids were ever destined for *Star Wars* obsession, they were mine.) They would raise their plastic weapons and, more often than I'd like, accidentally hit the pendant light hanging down from the plaster ceiling. The light would swing on its chain, looking as though it might fall down on their heads. Every time it happened, I thought: *I have to do something about that.* But I didn't know what to do or how to do it, so it became just another job left unfinished.

After one particularly cinematic duel, I decided that either a boy or the light was going to end up damaged. The boys were probably staying, so the light would have to go. I wasn't going to call anybody to take down my light, either. If I was going to be a leader, I had to take the lead.

It didn't seem all that hard, apart from the apparent guarantee of electric shock and a fall off the ladder. I spoke to one of the Widows and took detailed notes, and I did more research online. I wrote out my usual meticulous, step-by-step instructions and went over them again and again until I felt confident in my quest. If logic and evidence failed me, I hoped I'd get lucky.

I turned off the light. I got out the ladder. I climbed up, trying to ignore the way the floor creaked underneath me. I reached high over my head and took down the fixture— carefully unscrewing the necessary screws and more carefully disconnecting the black wire from the black wire and the white from the white. I climbed down the ladder and set down the light. Then I climbed back up the ladder. I took some electrical tape and wrapped it around the raw ends of the now-exposed wires, still dangling from the hole in the ceiling.

I climbed down the ladder and admired my handiwork. The foyer was a little dark now—you would have thought I'd understand what happens when a light goes out—but the boys wouldn't hit the pendant again, no matter how crazy their duels became. I had made the sort of small fix that could masquerade as triumph.

I felt so reduced. I felt so gigantic.

- - - - - - - - - - - - - - -

CHAPTER 17

Chance Encounters

Sitting on the small plane bound for Thunder Bay, Ontario, I wondered why I had ever said yes. A long weekend away from the boys, and right before the first in-person Starshade meeting—one of the most important meetings of my life.

I had agreed to talk at the Royal Astronomical Society of Canada's annual general assembly. Normally I would have declined such an invitation, summer weekends being precious to me, but the RASC had given me so much. The first time I had looked at the moon through a telescope—five years old, my father standing beside me, my eyes wide—had been at a society party. Later, as a teenager and through university, I had attended nearly every Toronto branch meeting. So although I was days from flying to NASA's Goddard Space Flight Center in Maryland to start trying to make the Starshade, I had a different commitment to meet.

While I was in my bedroom packing for the trip, Alex had lounged on my bed, watching me. I couldn't decide what to wear. "Sometimes it's hard to choose," I told him. "I don't want to be too formal, because it's supposed to be a fun gathering over a long weekend. But I don't want to be too casual, because then it will look like I didn't take the invitation seriously." Alex looked at my closet and then at the clothes in my hands and nodded. "I get it," he said. "Women have too many clothes, but not enough of the right clothes." I laughed and savored the moment. Worrying about what to wear meant that my other worries had faded, even if for only as long as it took for me to pack my suitcase. I was doing okay.

I stared out the plane window at the clouds below us. In a couple of weeks, it would be the two-year anniversary of Mike's death. The Widows had made no plans for it. One year removed from tragedy called for fistfuls of sparklers, but two years didn't call for much beyond a few sad hours of self-reflection. I had read somewhere that widows either meet someone new and remarry within two years or they never do. I couldn't help thinking that on that front, at least, I had run out of time.

There was a reception before my speech, the kickoff to the weekend-long event. We were at a small university emptied for break, where big windows looked out over rocks and trees. Thunder Bay presumably has that name for a reason, but it was sunny and warm that week, light until ten o'clock. Canada's long summer days always made up for its short winter ones, I thought. The resignation I'd felt on the plane fell away. There were nearly one hundred amateur astronomers in the room, excited to revel in their shared passions, and I felt the addictive spark of belief. Most of them knew that I was a guest speaker, and I was in constant conversation, talking mostly about the relative merits of one telescope over another.

I remember the instant I first saw him. I could actually feel my head turn toward him the way a sunflower follows the light. He was very tall, so he stood out even in the crowd, across the room. He had long hair gelled back away from his face, dark-rimmed glasses, and a wide smile. He was tan, as though he'd spent his life in the sun. He wore a crisp white shirt that made him look darker than he already was. *Wow,* I thought. *Who's that man?* I decided that I had to meet him. I didn't know how. I tried not to stare.

He stared back for a moment. Then he turned and glanced back over his shoulder, as though he thought I must have been looking at someone or something else behind him. Then he looked back at me, and our eyes connected. That look was all we shared. It felt like plenty.

When I got up on stage later that night, the tall man was sitting near the back of the auditorium. I was still trying not to stare. I pretended to concentrate on my notes, though I had my entire talk memorized. I had lost count of how many times I had given it. I talked about my search for the first Earth-like twin, my lifelong quest to see the smallest lights in the universe.

I was relieved when I was finished. My presentation had gone well enough, and now I would get to sit back and enjoy the rest of the lectures, picking up again the next day. Since astronomy conferences aren't exactly high-end, I was billeted at the home of a spunky amateur astronomer with bright silver hair. She was a widow. We talked a little bit about that, about dating again. She told me that she had been asked out several times but had always said no. I tried to give her some advice about men (she should go ahead and date when she was ready) and money (she should go ahead and renovate her bathroom). Then we turned to astronomy, my real area of expertise.

My hostess was a devout Catholic, but she also loved the stars. She knew that many in her religious community would have trouble navigating the discovery of other life in the universe and what that might say about their faith. Their fear was understandable. The history of their (and every) church includes its scientific challengers—Copernicus, Galileo, Darwin— and they would soon have to decide, again, what one of their belief systems meant for the other. It wouldn't be long before religion couldn't provide people with every answer anymore. We stayed up well past midnight discussing what we came to call the Awakening.

I probably slept more deeply than she did that night. The next day, I headed to lunch in the university cafeteria, a little later than almost everybody else. It was mostly empty— except for the tall man. We found ourselves standing together at the salad bar. I turned to him. I didn't know what to say, so I waited to see if he would say something.

He cleared his throat. Then he reached out his hand.

"Dr. Seager, my name is Charles Darrow," he said. There was a pause. Charles Darrow seemed nervous, but he also seemed resolved to say what he said next. "Would you like to have lunch with me?"

We took our trays to a quiet table on the far side of the cafeteria, next to the floor-to-ceiling windows, overlooking all of those rocks and trees. It felt like a kind of victory to have found my way to the same table: *I finally get to meet him*. I had never had such an instant attraction to a man. I looked right at Charles, wearing a giant smile. We told each other a little about ourselves, bare bones, first impressions. I needed to focus to listen to him. Charles was from Toronto, the president of my old branch of the Royal Astronomical Society of Canada. Ours really is a small world. He had been spending a lot of time up at his cottage in a place called Tiny, on the shores of Lake Huron's Georgian Bay. He sat on the beach during the day and looked through his telescope at night. I wondered why he was spending so much time in a place named Tiny, but I didn't ask, and Charles didn't say.

That night we met again at another talk, and Charles asked if he could sit next to me. We sat in the dark and listened to the speech and the sound of each other's breathing. I liked being next to him. He was so much bigger than me, I felt in some ways as though I were sitting next to a source of gravity.

The following morning, I waited at the airport for my early-morning flight home, a quick turnaround before I headed to the Starshade meeting at Goddard. Charles had given me his business card. I pulled it out and held it between my fingers for a little while. I wrote Charles an email, saying that it had been nice to meet him. If I'm being honest, I didn't think we'd cross paths again. He had his life, and I had mine. I looked up the distance from Toronto to Concord: There were 549 miles between us. We might as well have been living on different worlds.

·

I was home for a single afternoon, long enough to unpack and repack my suitcase, take the boys for a quick swim, and thank Jessica for looking after them. Back to the airport. Back on a plane. I read over the notes for my next talk, this one to an audience of professional astronomers rather than amateurs. It's always been interesting to me how different the same ideas can sound, depending on the audience.

"So, I hear you're an expert on the Starshade," I heard someone say, a little snidely, not long after I arrived. Someone else repeated the same refrain almost word for word. He, too, meant it sarcastically, not sincerely. He was an engineer, and I was a scientist, so I could see where he was coming from: He knew better how to build things. Still, I was upset that my supposed colleagues would be so rude. I knew that they wouldn't dare do that to a man in my position, with my qualifications, but I tried my best not to look stung.

I sat down with my STDT for the first time: There were ten of us on the committee, including a couple of emissaries from the Jet Propulsion Laboratory. That's where the Mars rovers and other fantastic machines are built. We also had an invaluable design team working alongside us. There was no point in dreaming up something that couldn't be made, and they would remind us of the limits of turning blueprints into physical objects.

That first day passed in a blur, but I liked our odds more and more as each hour raced by. Despite our wobbly start, I began to feel like we were part of a team. It was a well-assembled group, with complementary skills. By the time we sat down for dinner that night at a local Thai place, I was tired but happy. We were going to make something amazing.

On the second day of meetings, I woke up early to give an interview to the BBC for World UFO Day. Some people celebrate it on June 24, when an aviator named Kenneth Arnold made the first widely reported sighting of a UFO. While flying near Mount Rainier in Washington State in 1947, he claimed to have seen nine objects traveling in formation. He said that they were shaped like pie plates—which is why we refer to "flying saucers," and why fictional alien vessels are so often portrayed as disks. (I've always thought that was strange, because we've never built a flying disk ourselves.) July 2, when I was being interviewed, is an equally recognized World UFO Day. That's when aliens supposedly crashed into the New Mexican desert near Roswell, only a few days after Arnold's sighting. That was some summer.

I don't believe we were visited by aliens in 1947, or any

other year. I don't think anyone or anything can travel that far. But I still wanted the global audience of the BBC to imagine the possibilities. I put on my favorite brown leather jacket— I could hear the whispers of fashionable Chris, reminding me not to dress like a dork on TV—and sat in the early-morning light in my hotel room, waiting for a video call from a very proper-sounding British broadcaster named Tim Willcox. I prepared to make my standard argument.

"What is your hunch?" Willcox asked, and he laughed a little. "It's not very scientific, but what is your hunch: Do you think there is life out there?"

"Here's the thing," I said, smiling. "Scientists never like to say yes or no to something without any evidence whatsoever, especially on the BBC. I have to tell you, though, statistically speaking . . . If you just do the math and ask how many stars are out there, and how many planets are out there, it really seems inevitable that there is life elsewhere in our universe."

"Well, I'm hoping you find Krypton," Willcox said, "and find a new Superman as well. Sara Seager, best of luck, and talk to you when you've found them, perhaps."

I wanted so badly to give that follow-up interview. Can you imagine? The day that we find proof of other life in the universe? The day that we *know* someone else is out there will be one of those days that stands like a sentinel between the before and the after. I pictured the silver-haired widow in Thunder Bay turning on her radio and hearing that news, her hand to her mouth, tears in her eyes: the Awakening.

I returned to Goddard possessed. I gave my talk and listened to several others, about the best exoplanet candidates for life and the evolution of celestial giants. The room was soon crackling with creative energy. We used the word "targets" a lot. We were setting our sights.

I was at the airport waiting to fly home from Goddard when I had my first chance to check my email. I saw a return message from Charles. I nearly fell out of my chair.

There's something I forgot to tell you, he wrote. *Can we talk?*

•

On the Fourth of July, I got together with Melissa, our kids, and some of her non-widowed friends. I knew I was doing better because I could stand their giggly company without welling with contempt: *Oh, I don't want to murder them, that's good.* We took the Red Line to Cambridge, to the MIT campus, bound for the roof of the Green Building, twenty-one stories up. It was the perfect place to watch the Independence Day fireworks over the river. I showed Melissa my email exchange with Charles. By then I'd recovered my composure enough to reply to him as casually as I could: "Sure." Now I was waiting for my phone to ring.

"Fun!" Melissa said above the noise of the train, and she let out a peel of laughter. I knew what she was trying to tell me. *Just have a good time with this one. Don't take him too seriously.* For once I thought Melissa didn't have all the answers. I was shaken to the core, as though I'd found some new knowledge, left reeling by my own version of the Awakening. Charles and I had met over the course of one weekend in Thunder Bay and hadn't spoken since. I was a scientist with no evidence whatsoever. But I knew that there was something about him; he wasn't like the others. He wasn't like anyone else.

He called the next day. I found out later that he had picked up his phone five times before he finally found the courage to finish dialing my number. My heart beat faster when I saw Toronto's familiar area code on my screen.

"Hello?"

"Hi, Sara. It's Charles."

We didn't really talk about anything. Just small talk. *How are things? Good. Busy, but good.* We still didn't know each other very well, an accident of our shared awkwardness. I had told him that I was stressed about spending enough time with the boys, torn between them and my passion for the stars. Charles told me that he thought my work was important. He also told me that his father had been a traveling salesman before he started his own demanding business, and he grew to regret the time that he had spent away from his kids. Charles wanted me to know that he thought I was a good person and a good mother for worrying that I hadn't found the right balance. If I hadn't found it, I would.

There were glimpses of himself in what he had to say, but I felt as though there were important things about him that he wasn't telling me.

I didn't tell him some things, too.

We hung up, but I kept thinking about him, and not in my usual way. I wasn't analytical. I performed no dissections. I knew only that I wanted to talk to him some more.

The day after our call, the boys and I flew to Switzerland to visit Brice again. There was no work involved—a true summer vacation. On one rainy hotel-bound day, I decided to send Charles a short email, the gentlest of nudges. *Hi from Switzerland,* I wrote. *We're trapped in a thunderstorm. What's new with you?*

I was a little surprised when he soon wrote back.

Have some rösti for me.

I lay on my bed and laughed. Such an unexpected response. We wrote back and forth a couple more times, and Charles always wrote something that made me smile. He asked whether we could speak again when I got back home. *Of course,* I wrote. This time, though, I told him that I wanted to talk via Skype. I wanted to see his face. I wanted to see his eyes.

Okay, he wrote. *Let me get a camera for my ugly mug.*

Ugly mug? Whenever Charles looked in the mirror, he must not have seen what I saw. I thought he was beautiful. He looked at himself like his own worst audience. I looked at him and saw life itself.

•

Part of me wanted to tell him everything, but Charles and I didn't talk like that to each other. There was no urgency in our confessions. There was no race to the next step in our relationship, because there was no next step. We were becoming friends, and true friendships take time. We gave each other only thin clues about who we were and what we had gone through. I told myself that I was leaving the universe to its own devices, but that wasn't entirely true. It was more that I knew what I wanted the universe to do, and I didn't want to do anything to get in its way. I was certain that something bigger was coming for us. I was determined to let the inevitable happen.

I had told Charles that I had children, but he didn't know who their father was. He didn't know whether I was still married or divorced or any of the other possible options. He knew that I was an astronomer, and he knew that I taught at MIT, but he didn't know exactly what my life was like, or just how famous I had become in my field. He knew I liked him, I think, but I don't think he knew how much.

Charles told me that he worked at the family business in Toronto. I found it online. It was a wholesaler of specialized machine parts. He knew how tools worked. He didn't just know that ball bearings existed; he knew *why* ball bearings existed, and he knew that there were different ball bearings for different jobs. I admired that about him, the control that his work seemed to afford him over the world. He worked regular hours, had his weekends off, took vacations when he needed them. There seemed to be an order to his life. He had the measure of things.

Then Charles said something else. "My wife," I heard.

His wife. Present tense. No adjectives.

He was married.

That's okay, I thought in a self-protective rush. *I'm not ready for a relationship anyway. We're friends who flirt a little. Just a little crushing on someone from afar. No harm done.* But inside, I had sprung a leak and felt myself deflating. I told myself that I was being silly. *Five hundred and forty-nine miles? We never stood a chance.* I knew that. I *knew* that. Still. I had let my heart drift a little. I had allowed myself to feel a little hope.

Charles and I kept talking. We never ran out of things to say. Even if we talked about something ordinary, my attention never wavered. It didn't hurt that he was quick with his jokes. A lot of them were self-deprecating. He had a sardonic way of looking at himself, and he made me laugh more than I had laughed in years. I could be myself with him, and he seemed drawn to me, not repelled. Conversation with Charles felt so different from most of the time I'd tried to get to know someone. It felt as though he and I could talk forever.

I told Charles that Riccardo, the MIT alumnus who had kept his promise to be alongside me for the rest of my life, sometimes wrote to tell me good night and to wish me a good morning. He took the hint.

Charles wrote me that night: *Good night, Sara.*

I woke up the next morning: *Good morning, Sara.*

It was time to tell Charles my own secret. It was a few days before the second anniversary of Mike's death. I told him what the next week would bring.

He didn't apologize on cancer's behalf or try to find the silver lining in the death of someone you loved. He said that he hoped I would be okay. That he would be thinking of me.

I told him that it was going to be my birthday that weekend, too. How depressing.

I woke up on Saturday. *Good morning, Sara. Happy Birthday.* I wrote Charles back: *My birthday's tomorrow. Sunday.*

You just said "this weekend," he wrote. *I had a fifty-fifty chance.*

•

In early August, I had to go to Toronto to help close my father's estate. There were still loose ends. I asked Charles if he wanted to get together to give me a break from tying them up. He picked me up in his spotless white Volkswagen. I was excited to see him but soon felt a little confused. He seemed nervous, especially compared to how relaxed he was during our online conversations.

We had planned to visit the David Dunlap Observatory, north of the city. He wanted to take me on a tour. I joked with him that I should probably be the guide, since it was the same observatory in which I'd spent two summers back at the University of Toronto. I knew every inch of the place. It was built in the 1930s, a majestic dome manufactured in England and shipped to the less-polluted skies of the colonies. It was perched on the top of a hill, surrounded by a protective canopy of ancient trees. Inside there was a massive seventy-four-inch telescope, the second-largest in the world when it began scanning the sky. The school had since sold the property to a developer, who had cut the observatory from his holdings and provisionally given it to the Toronto branch of the Royal Astronomical Society of Canada. Which explained why Charles now had the keys.

"Pressure's on," I said.

I had told him that the library was my favorite room. "The library is off the tour," he teased.

At the observatory there were so many reminders of why I do what I do. I had learned so much back then. I had learned so much since. Despite our shared passion for the stars, Charles and I never spoke about its origins. We didn't need to. You don't need to explain your love for something to someone who loves the same thing. They already know your reasons.

We took our first picture together next to that giant telescope. Afterward, Charles took me to dinner. He gave me a present, a small box. I opened it, and inside I saw the glimmer from a watch. It wasn't a normal watch: it was an astronomer's watch, with a white band and a big face. It had a sundial—given a specific location, it showed when the sun would rise and set—and a moon dial, too. I would be able to look at my wrist in the afternoon and know the phase of the moon that night.

"Thanks for being my friend," Charles said. He said it in a way that made me think that he was plumbing a deep sadness, a sadness that, so far, had been off-limits to me. I had resigned myself to a long-distance friendship. Emails, Skype, texts, maybe a visit whenever I happened to be in Toronto. Two people who could love the stars together. Now he had given me a watch—a watch that would make me think of him only every time I wore it, only every time I wanted to know the where-abouts of the sun or the moon.

I was consumed with a single question: What were Charles and I?

We finished our meal and he drove me downtown, toward Lake Ontario, where I had a hotel room waiting for me near the harbor. I was flying out early the next morning from the Toronto Islands. We pulled up outside the hotel. Charles turned off the engine, and we sat together in the car gone still. The air was thick. I felt as though I were being crushed under an enormous weight.

I looked at my watch on my wrist and knew the phase of the moon and what I wanted to say. I wanted to send Charles a simple, clear message: *We belong together*. I didn't know how to deliver it. I didn't know how to begin crossing the gap between my theory and our reality, the gulf where so many good things go to die.

I could tell that Charles wanted to say something, too, but he was having even more trouble coming up with the right words. We both almost said something so many times: that last swallow before real confession; so many nervous licks of our lips . . . We sat there for seconds that felt like minutes, and minutes that felt like hours.

"Well," Charles said finally, and I rose up in my seat. "I have a long drive home."

That was my cue to get out. I said goodbye and lifted my-self out onto the sidewalk. It was cooler on the pavement in the middle of August than it had been inside that car. I took a deeper breath, and I watched him pull away.

CHAPTER 18

Clarity

"Sara," Bob said finally. "Do you know what I do when I need clarity? I run across the Grand Canyon."

It had been nearly two years since my tearful dinner with Bob Williams, and those words still echoed inside my head. There were many reasons why I couldn't shake them. I had traveled a lot since Mike had died, but I hadn't covered ground the way we once had, and I wanted to find the feeling we'd had when we'd paddled across Wollaston Lake: the intimidation of distance and the satisfaction of those who conquered it. With the Starshade on the horizon, I found myself drawn more and more to finite goals, to *accomplishment*. I also wanted to feel capable of achieving what I decided to achieve, to feel that my future was mine to decide. I needed to tax my muscles and test my resolve. The Grand Canyon was my kind of challenge. It was adventure as metaphor. There was a rim for each of my lives, divided by a chasm that was within me to cross.

I also recalled Bob's concern when he thought I might actually make an attempt. It was clear from my reaction that I was probably going to try. He called me the day after our meeting.

"Sara, this isn't easy," he said. "It's called the Grand Canyon for a reason. Whatever you do, don't try to cross it in a single day." It was a real ambition but an easy one to shelve.

In the late summer of 2013, the boys had a couple of unscheduled weeks between the end of camp and the beginning of school. I was still fixated on the shortness of life. I asked Alex if he still wanted to make the 14,000-foot climb that he had once talked about so urgently. "Yes," he said. I asked him to rate his desire on a scale of one to ten. "Seven or eight," he said. The way he said it, soft and uncertain, made me think that his heart held a different answer.

Alex was growing up so fast. He was eight years old by then. For about two years, he had seemed stalled in his growth, a little coiled bundle of muscle. Suddenly he had grown several inches and gained thirty pounds. The greater the mass, the harder it is to move. Maybe mountains had become taller in his mind than they once were.

I wondered whether seeing an actual mountain would ignite some of his old feelings. I decided to take the boys to Colorado. Jessica would come along, radiating her usual enthusiasm, and I also bought a ticket for a young woman named Zsuzsa, a recent MIT graduate and hiker who had become another member of our extended family. I was still adopting stray adult humans the way I'd once taken in homeless cats. Zsuzsa needed a place to stay; I had empty beds in empty rooms, and I could always use more help around the house. She was also a character, claiming that her Russian father had trained her as though she might become an agent for the KGB. She said that she was built for danger. I had started to worry that our house would burn down in the middle of the night, and she insisted that she would save us.

On our flight to Denver I looked out the window, down at

the Rockies, and realized that we didn't have anything close to a plan. I hadn't even picked the mountain that Alex and I might climb. I looked down at their peaks and wondered: *Which one of them will be ours?*

It didn't take Alex long, after we'd made our brief survey of Colorado's fortress-like summits, to decide against making a climb. He didn't need to break world records anymore. Maybe that was part of his growing up, the distance widening between who he was and who he wanted to be. Whatever his reasons, I told him that was fine. I'd told Jessica and Zsuzsa in advance that we'd probably need a backup plan, and now we employed it. We'd drive to the Grand Canyon instead.

We got as far as Grand Junction, Colorado, and bunked at a hotel. There was a pool and a small amusement park not far away. Max and Alex seemed done with sitting in the car. Over dinner, we decided that Jessica would stay in Grand Junction with the boys. Zsuzsa and I would keep driving southwest.

"Sara," Bob said finally. "Do you know what I do when I need clarity? I run across the Grand Canyon."

Zsuzsa did the driving. It was a spectacular trip out of Colorado, through Utah, and into Arizona. Earth is the most wonderful planet. We drove down from the mountains and across river gorges into the first fingers of desert. Trees became cacti. Green became brown; brown became red. The temperature, the light, the smell, the sounds, the feel of the wind: With every mile something about our surroundings changed. It didn't feel like we'd crossed only a couple of state lines. It felt as though we had gone to sleep in one universe and woken up in another.

I called ahead to a hotel on the North Rim and asked if they had any rooms. One was left. A sign.

We pulled into the Grand Canyon Lodge, and it was one of

the most gorgeous places I could imagine. The rooms feel like log cabins. Outside, the hotel clings to the edge of a steep cliff, the sky and clouds behind it turning every color. When night fell I found a quiet socket in the rocks and looked up. There is nothing like the desert to draw out the stars. I saw the constellations through the eyes of an ancient Greek: the stars an eternal mystery, the stars come to life.

Zsuzsa and I had made the barest of outlines, but I was glad for our lack of preparation. I might have become anxious given more time to contemplate the details. We agreed that we would wake up early, tie on our boots, and shoulder our little backpacks. We'd carry water and sunscreen and snacks. If we needed more water, we'd drink from the taps at the springs that we would find on our way. There was a river down there, after all. We'd cross from north to south, spend the night at another hotel on the South Rim, and then make our triumphant return the next day.

We woke up in the dark and made our way to the canyon's edge. I thought we were about to do such a special thing, but we were far from alone. Close to a crowd had gathered at the starting line. I was disappointed but maybe also a little relieved. We were about to do something extraordinary that seemingly ordinary people could do.

The sun broke over the horizon. I felt a little burst of adrenaline inside my chest. I nodded at Zsuzsa, and she nodded at me. We weren't going to heed Bob's warning. We'd hike the canyon in a single day. We took our first tentative steps into the canyon. Gravity soon began pulling us toward the bottom, like water finding the drain. Hours into our hike, Zsuzsa found the energy to break into a run. I followed. My feet began to blur beneath me. I felt sweat bead on my forehead, the good kind of pressure in my lungs. Zsuzsa let out a howl.

"This is the best day of my life," she shouted, deep into the canyon.

•

It's impossible to look at its walls and not think about the passage of time. Geologists disagree about how long the Colorado River has been making its deepening, widening cut into the Earth's surface, but the best guess is something like five or six million years. Today the canyon is 277 river miles long, and 18 miles across at its widest point. It is, at its essence, a very big hole, carved out grain by grain by wind and water running over sand and rock. It's a testament to how monuments get built: little by little. The canyon's walls, a mile high in places, have left about two billion years of geologic history exposed.

With every few steps we took down from the North Rim, we leapt across centuries. There went my lifetime. There went the Spanish Civil War. There went the Renaissance. There went the Puebloans, and there went the Holy Roman Empire, and there went the Shang Dynasty. There went the origin of our species. There went the dinosaurs, and prehistoric reptiles, and the beginnings of life itself. Give the Colorado River enough time, and it might unearth the Big Bang.

Zsuzsa and I also measured time in more immediate ways, most keenly as distance. The hike from the North Rim to the South is twenty-four switchbacking miles. With a miles-long descent. And somewhere over there, across the river, as long of a climb. We knew that the down is deceptive. It's halfway, but it isn't. Down is easier than up.

We also measured time by daylight. We were racing the

sun. That day was overcast, which was good as far as temperature was concerned. It was less ideal when it came to how far and how late we would be able to see.

There is a bridge across the river at the canyon bottom. I looked out at the water on its endless course. It had been more than a decade since Mike and I had run that water, since I had waved up at the people who had watched us from that same bridge. I had a moment of sadness, the shadow of an ache. But it didn't last long. I went to a different place. In some strange way, I could almost see my own ghost there, my younger self in the water, the years falling away. I knew everything that had happened to her, as though she were a stranger whose fortune I could read. I could see her in the river and call out and tell her. Should I tell her?

The clatter and clomp of a mule train, carrying tourists to the bottom, broke my weird spell. They were led by an older man in cowboy clothes. *He's handsome,* I thought, and then I found myself thinking about the thought. *He's handsome.* That thought was a simple, involuntary reflex, a conditioned response to positive visual stimulus. That thought felt like its own kind of bridge. I might finally be able to cross from one side to the other.

Zsuzsa and I were practically giddy. The three-quarter mark, the most desperate point in most journeys—so much ground covered, so much to go—passed in a cascade of good feeling. My legs felt great. My spirit soared. Zsuzsa and I talked about maybe even making our return that night. The clouds were clearing, and there was going to be a full moon.

Then her knee began to bother her. She told me to go ahead. I made the climb out of the canyon alone. I sat on the rocks and looked out at its red walls in the last of the golden light. Just when I decided I should go back down after her,

Zsuzsa finally climbed out, too. We had made it to the South Rim. It was crowded with tourists; I remembered Mount Washington, the drivers and the hikers and the difference between them. I would sleep in my clothes. I would sleep canyon deep.

That evening, Zsuzsa decided that she couldn't make the return. Her knee wasn't up for it. She was disappointed that she'd have to take the long bus ride around.

I'd walk alone. Early the next morning I found the start of the trail back down and waited in the dark for the dawn. I started with the sun. I ran down into the canyon and reached the bridge at the bottom. The handsome old cowboy was there again.

"I saw you yesterday," I said.

His white teeth shone out from under the brim of his hat.

"I love you," he said.

It was a surreal moment. He said it so easily. I didn't feel like he was hitting on me; I felt like he was playing things up for the tourists around him. But he sounded sincere, as though he were stating another geological fact for them. I smiled the widest smile. A cowboy's love and my heavy legs lifted me from the bottom of the canyon and into the climb. I wiped the sweat out of my eyes and the dust out of the corners of my mouth. I tried not to look up. I focused on the rocks in front of me, layer after layer. Here come the dinosaurs. Here comes the Spanish Civil War.

Zsuzsa was waiting for me on the other side. I took a long shower, and we sat down together for the most satisfying meal. I felt electric. I felt ecstatic.

The next morning, we began our drive back to Grand Junction. Zsuzsa took the wheel and I eased my seat back, my body heavy with lactic acid and pride. I watched the desert turn back into forests, canyons on their way to becoming mountains. I wasn't sure that I'd be able to explain to Jessica and the boys what the trip was like, or how that journey had allowed me to know exactly who I was, exactly then.

"Sara," Bob said finally. *"Do you know what I do when I need clarity? I run across the Grand Canyon. In a single day."*

I called Bob as soon as I had reception. Although we spoke once in a while, I hadn't even let him know I was going.

"Hi, Bob," I said. "I made it."

I'd made it, and that's when I knew. It had never been a problem of distance. It was only a matter of time.

CHAPTER 19

Flashes of Genius

In science, sometimes your hunt for one thing leads you, or someone else, to something better. At our best, scientists are explorers.

I could still feel the Grand Canyon in my legs when school started up again that September; I felt strong in a way that I hadn't for a long time. As soon as I'd returned from the trip, Charles and I had started our constant communication again—email, text, Skype, phone—picking up as though we'd never spent that awkward eternity in his Volkswagen. We had settled on being pen pals, and I decided that was for the best. Instead I poured my heart into the Starshade, prepared to return to teaching, and plunged back into my favorite research. I vowed never to think that we'd reached the edges of our maps.

Kepler had been busily charting its thin slice of the galaxy, a swath of sky in the constellations of Cygnus and Lyra, and delivering a steady stream of hopeful discoveries. Fourteen previously undiscovered exoplanets had been confirmed at the end of August, bringing us to about 150 named new Kepler worlds, with thousands more candidate planets waiting in line for validation. The volume of discovery allowed our community to start finding patterns in planetary systems; we were able to conduct our version of a census. It was more than stamp collecting after all. Astrophysicists who were interested in how planets form, for instance, now had so many more examples to study.

I left that work to others. I wanted to push deeper into the new frontier. My research into biosignature gases with William Bains had taken a surprising new turn. I thought about which gases other than the obvious ones—oxygen, methane— we might see as a sign of life, and I was struck by how many gases are the products of biology. With the exception of noble gases like helium, which are inert and nonreactive, I wondered whether every single gas might be produced by life. I pitched my theory to William, who immediately dismissed it as ridiculous. But I countered: Every gas that exists in Earth's atmosphere, even those detectable in parts per trillion, can be made by living things. They usually have nonbiological sources, but life *can* make them. William decided my notion was slightly less ridiculous than he'd first thought, and along with a brilliant postdoc named Janusz Petkowski, we put considerable work into proving my crazy idea. We decided to head over to Harvard one afternoon to present our case to the biologist Jack Szostak, a Nobel laureate. Jack listened politely to our PowerPoint presentation, and then named one gas, off the top of his prizewinning head, that is not the exhaust of life.

If Jack could name one gas so easily, there must have been more; either way, one was enough to explode our theory. William, Janusz, and I were dispirited but not defeated, and we went back to work. We soon came up with a staggering result.

Given Earth's surface temperature and atmospheric pressure, more than fourteen thousand molecules can exist as gases, and a quarter of them are produced by life on Earth. Who knew which of them might be produced by life on another world? That confirmed how careful we needed to be in our search. Finding an alien atmosphere rich in oxygen would be a groundbreaking discovery. But it's far from the only possible sign of life. Oxygen is one of thousands.

Our rapidly expanding database was criticized by many of the other scientists working on biosignature gases. They were adamant that only oxygen, methane, ozone, and a handful of other gases would ever be produced in large enough quantities for us to detect. I didn't care. I had worked hard to permit myself a more expansive definition of life. I knew that our work would serve as the foundation for more research, the launchpad for who knew how many other discoveries, for countless dreams.

Janusz had already unearthed something amazing on his own. He'd looked at those gases and solids that aren't the products of biology and found specific, unmistakable patterns in them. There are fragments of molecules—we named them "motifs"—that life largely refuses to make. We had the feeling that motifs, those biochemical voids, were as filled with possibility as an unexplored ocean. Nobody knew what we might find in them.

For instance, if we look at the hundreds of thousands of different molecules that we know life does create, about 25 percent of them contain nitrogen. About 3 percent contain sulfur. Those are two fairly abundant elements in our world. But it's extremely rare to find molecules with nitrogen-sulfur bonds. That surprised us, because nitrogen-sulfur bonds are common in industry and pharmaceuticals; we force that union all the time in rubber manufacturing and for a variety of dyes and glues. But life? Life almost always avoids forging those bonds, and when it does, they're often toxic.

Instead, life often bonds sulfur with hydrogen. Proteins containing that compound are ubiquitous; they likely exist in every cell of every organism on Earth. Together, the three of us theorized that nitrogen-sulfur compounds and sulfur-hydrogen compounds can't easily exist in each other's presence in nature. Life on Earth has chosen hydrogen as one of sulfur's most common partners, and nitrogen will rarely be invited to join. They are close to mutually exclusive.

Does that mean that if we find an exoplanet with nitrogen-sulfur compounds in its atmosphere, it must not harbor life? Not necessarily. Some of my peers might argue that ignoring the lessons of life on Earth will make the parameters of our search impossibly broad. I disagree. If my research into bio-signature gases has taught me anything—and it has taught me many things—it's that life will find its own way, and it isn't always the same way twice. We need to think more radically, not less.

So let's imagine.

Imagine for a moment that we somehow visited that strange planet bathed in nitrogen-sulfur compounds. We would land our rocket on its surface. Whatever life might be there, especially if it's intelligent life, would presumably gather to meet us. We would crack open our hatch, step out onto that alien surface, and reach out our trembling hands. And the sulfur-hydrogen compounds in our bodies would leach through their skin and contaminate the nitrogen-sulfur compounds in their bodies, and the reverse would happen to us. We'd poison them, and they would poison us, and everybody and everything would begin a slow, deadly march toward an apocalypse of chemistry.

The same would hold true if those aliens came here. If they ever open that door, life as we know it wouldn't just change. It would disappear. Earth would begin again. And something new would take our place. Life would find another way.

•

I was on the train home from MIT, doing my version of staring into space, thumbing through my cluttered in-box, when I saw an email from the MacArthur Foundation. Seeing that name gave my heart a little flutter. The MacArthur Foundation gives out the famous so-called "genius" grants: $625,000 over five years, no strings attached, given to people for doing inspired and inspiring work in every conceivable field.

The email said they had tried to call me earlier that day, but my protective assistant had refused to put them through. I supposed that they hadn't offered the MacArthur name, and he had assumed it was an especially trumped-up crank, vast networks of which now reached my office weekly. It was as though every new planet brought with it another conspiracy theory, and I had become famous enough to become one of their many targets. I couldn't be mad at my assistant. He was doing what he thought was right, and I was grateful that he was so protective of me. Still: Not being told about a call from the MacArthur Foundation! I can't imagine that had happened very often.

I wrote back apologetically and asked them to try calling me again the next day. That left me with a long night of wondering to endure. What could they want? A year or two before, they'd called me to check the references of an acquaintance they were planning to honor. Maybe they wanted me to help vet someone again? That was probably it. They needed a source, not a subject. But as I watched the trees streak past the train's windows as though they had lifted out of the ground and broken into a run, I couldn't resist the thought: *Maybe it's my turn.*

The next day they called again. This time I picked up. They asked if I was sitting down, which I was, though I had already started floating out of my chair.

It was my turn.

Sometimes in life, something momentous happens to you, but you realize the enormity of its impact only in retrospect. It takes time for you to understand its significance, how some seemingly innocuous choice or unforeseen event made all the difference for you or for someone you love. The phone call from the MacArthur Foundation wasn't like that. That phone call was one of those rare times when I knew something life-changing was happening in the moment that it was happening. I could hear people in the hall outside my office who had no idea what was taking place on my side of the door, but I knew. In that moment, I almost stepped outside myself, the object of an experiment: Watch what happens when someone's world takes a turn.

"Now, Dr. Seager, this is a secret," the people on the phone told me. I had to keep the news of my win to myself until the public announcement of the awards, scheduled for later in September, a little more than three weeks away.

"You can tell one person," they said.

They meant it as a consolation, as a balm: *We know this is a secret that's too big to keep, so here is our gift to you.* Except that it didn't feel like a gift. I could tell one person, because I was supposed to have found my one. I had. But then I had lost him. I'm not sure that I've ever felt Mike's absence more deeply than I did then, in a moment that had been made of the most delicate, perfect glass. I had received some of the best news of my life, and the second I hung up the phone—"Thank you, thank you so much!"—I was wracked by sobs. I remembered again the promise I had made to Mike when everything but our needs and wants seemed in short supply: *We'll have time someday. We'll have money someday. We'll have time and money someday.*

Now I had money, at least. But you can't keep a promise to a husband who isn't here anymore. You either keep it when he's alive, or you don't.

I knew instinctively, the way you know that a cold is going to be a bad one, that I was about to suffer one of the deeper strains of sadness. It wouldn't just come and go on its own; I would have to fight it if I wanted it to leave. I had asked if I could tell *two* people, and the MacArthur Foundation was kind enough to say yes. I was thinking of Max and Alex, but then I changed my mind. Why would they need to know in advance that their mother had won an award they knew nothing about? Instead, I told Melissa. She roared like a lion.

And I told Charles. He was surprised that I'd share such a big secret with him when we were still keeping so many.

The day before the announcement, I also told Marc Kastner, the MIT dean who had been so supportive of me when I was faltering. The money that he had found had paid for more hours for my small army of help. Each hour they worked was another hour for me to do my work. I went to his office and thanked him for every one of those hours. He surprised me by giving me a great big hug. I'm not the most huggable person, but he was ecstatic for me and I think for MIT, too. Only seven years before, there was such skepticism about whether I should have been hired. More recently, I'd been in his office talking about quitting. I was that overwhelmed little girl standing by a lake in the nighttime. Now I was about to be called a genius for everything I had learned since.

•

I woke up the morning of the MacArthur announcement with a strange feeling in my stomach. Things felt almost anti-climactic. The reality hadn't sunk in entirely, but I'd had weeks to process what the award meant for me. I had decided not to spend the money on anything dramatic, other than a big gift for the boys and a few more trips. Being a working single parent is expensive, and even with Marc's help, I had plundered most of my savings, forging ahead without any real plan or hope for stability. Now I'd been thrown a financial life ring. I would put most of the grant toward childcare and groceries and help around the house. I still did some chores—I even liked doing laundry, with its stark before and after—but I had learned by then that there were vast swaths of human existence that I would never master, and that was okay. There were other things at which I was one of the best. It would be a relief that I could finally talk about the MacArthur—that secret proved really hard to keep—but mostly that day was about the outside reaction to the honor, not my own. I already knew how I felt.

The moment of the announcement was like hearing a knock at your door and opening it to find a parade. MIT sent out press releases; colleagues and students stopped by my office to congratulate me; Max and Alex were with Diana in a pizza joint and saw me on TV; my phone chimed constantly and my in-box filled up until it overflowed. The attention was flattering, of course, and everyone was very kind. I still felt a little hollow.

In the chaos, I gave a phone interview with *The Globe and Mail*, a national newspaper in Canada. I wasn't in the right space to sound as magnanimous as I should have. The reporter asked me why I had left Canada to come to the United States, and I was surprised enough by her question to answer frankly: "Because America is better at fostering greatness." I tried to catch myself, realizing how that might sound back home: "But don't forget to put in that I still love Canada," I said. She included the entire quote, which made me sound less than genuine. I really do love Canada; it had just offered me different gifts than America had.

We also talked about my widowhood, and I told her how the MacArthur people had given me permission to tell two people. I'd told my two best friends, I said. I hoped that Charles would read the story and connect the dots. I was stuck in the middle of one of the hardest lessons of widowhood: It's the happiest moments when you feel the most alone. I was desperate for that day to have some love in it.

Then there they were, just waiting for me to hang up the phone—the Widows of Concord, cheering and clapping and smothering me with hugs. My legs almost gave out. I'm still not sure how Melissa managed to corral everybody into my office, with just enough room for their wide smiles and open arms, a picnic lunch and bottles of chilled champagne. We sat down at my long wooden office table and dived into a feast. I basked in their warmth. I was sitting next to six suns.

The last time so many of the Widows had been in my office, it was back when I was thinking of quitting. Now they were popping champagne corks into my ceiling, some of the only true friends I had made in my entire life, telling me that they were proud of me, they were happy for me: "What are you going to do with all that money? Sara!"

Winning the MacArthur was a blessing. It would do for me exactly what it was designed to do. It would encourage me to keep going in an almost literal sense: It would put courage into me. It would bring me the privilege of focus. But the Widows were the fellowship that mattered most to me. Underneath our celebratory surface, we all knew the real reason they were there. Nobody had to say it. I looked around that room and felt the day's first hint of a genuine smile on my face. Not because I wasn't sad anymore, but because I would never again be alone in my sadness. I held up my empty glass and waited for it to be filled. I knew that it would be. The Widows had always known what I needed before I did.

.

At the end of the month, NASA announced its latest exoplanet finding, this one by Brice the Swiss and other members of my research team. I had coauthored their paper. For three years, the Spitzer and Kepler telescopes had both been trained on the same mysterious world: Kepler-7b. Back in 2010, the planet had been one of Kepler's earliest finds, a hot giant one and a half times larger than Jupiter, orbiting so close to its star, Kepler-7, that its "year" is five days long. By way of orientation, Kepler-7 is part of the constellation Lyra; its better-known neighbor, Vega, is one of the brightest stars in the Northern Hemisphere.

Kepler-7b had been confounding from the beginning. It was brighter in its western half than its eastern one, but it was hard to know why. Perhaps Kepler-7b somehow had its own source of heat and light. Or maybe there was some other explanation for the imbalance. That's when Spitzer took over from Kepler. Fixing its infrared gaze, it helped determine that Kepler-7b was scorching hot, reaching temperatures perhaps as high as 1,800 degrees Fahrenheit—but it wasn't as hot as it should be, given its proximity to its star. It was finally inferred that the planet's western half was enveloped in a protective layer of clouds. They were reflecting the heat from Kepler-7, the way our atmosphere offers us relief from the sun.

An artist then painted our first portrait of Kepler-7b: dark and banded in its east, and wrapped in green clouds in its west. Kepler-7b was too hot to sustain life, but we had seen a glimpse of its face. Only a few centuries ago, we were drawing dragons on our maps to mark the ends of our oceans. Now we had divined something about the weather on a planet that orbits a star in a constellation the ancient Greeks had thought looked like a harp.

There are times when our progress can seem impossible, especially when we remember we're the same species that murders each other for oil and fills our oceans with plastic. But it's important that we take the time to appreciate how far we have come. It makes it easier to believe that we might go farther still.

When the news about Kepler-7b was revealed I was on a flight to Hawaii, soaring in more ways than one.

A tall man named Charles was in the seat beside me.

•

Charles and I were in different versions of the same unwinnable fight. He had a wife, but they were not happy and were in the middle of a long and painful separation. It had taken a miserable amount of time for their marriage to unwind; he had been sleeping on the couch for five years or so. She had told him what a failure he was for so long that he looked in the mirror and mistook beauty for hopelessness. He woke up and went to the same job he'd always had, one of two sons in a business of father and sons, stuck in the same traffic, trapped at the same intersections. His only escapes were those days he spent in the sun in Tiny and those nights when he looked up at his giants, the stars. Now he was about to turn fifty. "What am I supposed to do?" Charles said.

I was flush with my MacArthur grant money and the hubris that the mantle of genius can bring. Charles's fiftieth birthday was the first day of October. I invited him to come to Hawaii with me. Max and Alex could stay home with Rachel, their fun and obliging aunt from Alberta, and I could do some work and give a talk to justify the time away a little. Charles and I could stay on Mauna Kea, and I could call in a favor from a friend for a special tour of its telescopes. Whether Charles knew it or not, he had given me so much. He had shown me that I deserved happiness, and that it was possible for me to be happy again. It was my turn to give him something in return.

Charles said yes, and soon we were sitting beside each other on a plane over the Pacific. We had taken to sending the same text every time one of us had boarded a plane on our own: *If my plane crashes and I don't make it, I always wanted to tell you . . .* We would never finish the thought, leaving something important unspoken. Then, after we landed, we would text again: *My plane didn't crash, so I guess I don't have to tell you.* It had always made me smile. This was the first flight in months for which I hadn't needed to send the text.

We were still leaving something important unspoken. We were going as friends: separate rooms. I told myself that—*friends, we're only friends*—and Charles never offered anything that seemed like a correction. I was still trembling with excitement. We could talk to each other without instruments, if only for a little while.

We stayed in an observatory dormitory that's 9,000 feet above sea level. We weren't in the Hawaii of postcards. It was cold and hard to breathe. I'm always a little dreamier at high altitudes, and on that trip I surrendered completely to the gauzy spell of mountaintops. Charles and I visited my astronomer friend and toured the telescopes, assemblies of some of the most reflective glass on Earth.

There is an atmospheric phenomenon that is so rare, some people think it's a myth. I had never seen it myself, but not being able to see something doesn't mean it's not real. It's called the Green Flash. When the conditions are exactly right at sunset—a dead-flat horizon, a pollution-free sky, a sun that appears white-hot rather than red—the last of the sunlight, refracted through the atmosphere and around the curve of the Earth, will appear, for the briefest of moments, unmistakably green.

It's hard to see the Green Flash for the first time on your own, which helps explain why I never had. It's elusive in part because people want so badly to see it, but don't know what they're looking for. If you stare at the sun, even in its setting, your eyes will go blind to the sight. You need a partner, someone who is willing to sacrifice their own chance of seeing it on your behalf. You need to turn your back to the sun and have someone watch its descent for you, and then, at just the right moment, the instant before the sun disappears, your sacrificial someone has to tell you: *Now. Open your eyes.*

I was determined to see the Green Flash in Hawaii with Charles. I had announced my intentions to him long before. On our first night there, we drove to the Mauna Kea summit with a clear view of the ocean and the uninterrupted horizon beyond. The sun was white. Charles was almost visibly wilt-ing under the pressure; he, too, wanted me to see it. But there is no forcing the Green Flash. You have to wait for it to come to you.

We stood on the mountain. It was unusually cold, and we were bundled up against the wind. I turned my back and closed my eyes to one of the most gorgeous sunsets of my life. Charles faced the sun. I waited for him to give me the word. It felt like a very long wait, and I struggled to keep my eyes shut. I could feel him beside me, waiting, waiting, waiting.

"Now," he said. I turned around and opened my eyes.

And there it was: The Green Flash filled my watering eyes—an emerald green, perfect and pure. I smiled at Charles and he smiled at me, each of us overcome with something like relief. I felt as though in a moment I had been given a new horizon.

•

November. Every year in Concord, the bright blue skies and brilliant fall leaves give way to cold and rain and a thousand shades of gray. It's when I'm at my most melancholic, dark-eyed and prone to brooding. I finally acknowledged that I had given my heart to Charles and would never be satisfied with his friendship alone. Even though I wasn't privy to the thorn-ier details of his separation, I was worried that I was expect-ing too much, too soon. He had been going to Tiny nearly every weekend between spring and fall since the day he was born. He had worked at the family business since he was a teenager. His presidency at the RASC was meaningful to him, and his best friends were part of the same club. If he was going to be with me, he would have to leave the rest of his life be-hind.

I looked at the low clouds and empty branches and realized that I had made a terrible mistake: I had fallen in love with a man who was out of my reach. It wasn't my fault, really—love is one of the only blameless things we do—but I had spent enough of my professional life on roads to nowhere. I should have recognized when I was hurtling toward another dead end. One rainy November morning, Charles and I talked via Skype. We couldn't be friends anymore. Whatever our relationship was, we agreed to call it off.

I was devastated. I had been working from home that morning, but now I had to drive to the Boston NPR studio for a radio interview. I called Melissa from the car. Life had conspired against me again. I was a mess. I could barely see the traffic lights. I'm sure that she wanted to remind me of her Fourth of July warning not to take men so seriously, and to take Charles even less seriously than most, but she didn't. She took my side like a best friend should. "Life is messy," she said. "Things take time." She told me that even though life hadn't worked out with Charles, everything would still be okay. I told her that, to make things worse, I was driving to an interview on NPR. In the studio. How would I get through it? "At least it's not on TV," Melissa said. I showed up still crying, my eyes swollen and red. The host and technicians looked aghast. "Don't worry," I said. "No one's dying."

I still felt a terrible burden on my shoulders. Why was I so upset? I couldn't understand my own feelings. Why? We hadn't even kissed. Really, for once, no one was dying.

Later that month, I was asked out on a couple of dates by other men and went on them. One was with someone who had gone to the same Montessori school in Toronto that I had. He was coming to a conference near Boston, and he drove through a lashing rainstorm to take me out for dinner.

It was more of a reunion than a date. It was still me and a man eating together in a restaurant, so it was an evening of statistical significance. He was cute, smart, and incredibly sweet, with two daughters about the same ages as Max and Alex. We had a lot in common, including our childhoods. We must have passed each other in school hallways many times. I liked that about us, and I really liked him.

It didn't matter. Despite all of his excellent qualities—I'm surprised that the Widows didn't show up behind me and start screaming, "What are you waiting for?!"—the same thought kept tripping me up like an uneven step: *He's not Charles.*

.

In early December, I had to testify before Congress about the search for life in the universe. I would join two other experts in front of the House Committee on Science, Space, and Technology: Dr. Mary Voytek of NASA and Dr. Steven Dick from the Library of Congress. We would present our case on behalf of aliens and then take questions. Somehow I needed to make the case for hope.

Lamar Smith, the head of the committee, called the hearing to order. Mary wisely opened the meeting with an inspiring update. As of that late-fall afternoon, we had found more than three thousand likely exoplanets. The day before, Hubble had reported traces of water vapor in five giant exoplanet atmospheres. Water in the skies of giants isn't a sign of life, but it was its own kind of progress. There was possibility everywhere we looked.

One of the congressional members was Ralph Hall, a ninety-year-old Texan who had started his political life as a Democrat before coming to call himself a Republican. He was charming in his old-southerner kind of way. He looked at the three of us behind the witness table and said that we might represent the largest concentration of brainpower he had seen. "I just don't know how I'm going to tell my barber, or folks from my hometown, about your testimony here," he said.

We tried to keep our message simple. We needed continued support. We needed to make sure that children who wanted to become scientists were given every opportunity to see their aspirations come true. Most of my message was about the need to invest in more and better space telescopes, and about the value of the Starshade.

Ralph Hall stopped me. "Do you think there's life out there?"

"Do the math," I said.

Hall said that he couldn't do the math. That was the problem.

He asked us again: "Do you think there's life out there?"

"Yes," Mary said.

"Yes," Steven said.

"Yes," I said.

•

A few weeks later, I was headed to Guatemala to teach a weeklong workshop for astronomy students from across Central America. Jessica and Veronica would look after the boys. Alex had once said happily, while he, Veronica, and I were watching the Red Sox in the World Series on TV, "It's like we have four moms." (He was also including Diana.) I was a little taken aback at the time, because no mother wants to be one of many mothers. But I was grateful that my boys felt love from so many sources. More love was never bad.

I was still scared to leave them. My fear of flying remained unshakable. Charles and I had been engaged in complete radio silence, but before my flight to Guatemala took off, I decided to write him our old text. *If my plane crashes and I don't make it, I always wanted to tell you . . .*

I looked at it for a few seconds before I hit send.

I was shocked when I came home from Guatemala to find that Charles had written me a long email. He knew what I had known: He knew that if he pursued me, that if we pursued us, that he would have an old life, and he would have a new one. He had agreed that we shouldn't be friends, or anything more than friends, because he'd been overwhelmed by the prospect of leaping the canyon between those lives. Now he had made up his mind to try. Turning fifty didn't mean that it was too late for him; it meant that he didn't have any more time to waste. He had finished separating from his wife and moved into a basement apartment at his brother's house. He had started talking to his father about leaving the business. ("Congratulations," his father would say later, and he meant it.) The astronomical society could always find a new president. Somebody else could take every last one of his places. He had a new one to occupy.

I don't want to die unhappy, he said.

Many widows and widowers learn to protect their heart from more damage. They know that they can't take another blow, so they keep their love in an iron box, locked away inside their chests. Some significant percentage of them never date again, or if they do date, they date casually, without expectation or commitment. Maybe that's how their subconscious protects them from anything that resembles further loss. If they never love again, their hearts will never again be returned to them in pieces. One of Melissa's boyfriends had called her "guarded," and she was surprised enough to call me to talk about it. She rarely came to me for advice, and I thought carefully about what I should say. I told her that "guarded" isn't a word I would use to describe her. She's open and available, loving and committed. She's a source of light and heat.

I thought more about Melissa after I hung up the phone. As much as I loved her, we were different in fundamental ways. All the Widows were; all widows are. We respond differently to the same traumas. Nobody is wrong. One of the few traits the Widows shared was our honesty with the world, and we were all honest in our emotions. They just weren't always the same emotions, and we didn't always express them the same way.

I had decided once that I was going to be alone forever. My boys and I would never be kicked out of our Widows club. I could never judge anyone for thinking the same thing. But I didn't want to think that way anymore. I had always believed that great reward demanded great risk. My father had taught me that. Mike and the wild rivers and lakes of Northern Canada had taught me that. My boys had taught me that. Mount Washington and the Grand Canyon had taught me that.

More than anything else, space had taught me that. The stars had. Miracles don't happen in a vacuum. They are willed into existence by willful people. My losses had sometimes clouded my belief, in myself in particular, but now my eyes were clear and my lungs were full. For the rest of my life, I would rather suffer than experience nothing. My father had told me, so many years ago, not to rely on any man. He had told me that only a father's love is limitless. But how would I know how big romantic love could be if I never gave it another chance? I would choose for my heart to be broken rather than never feel a change in its beat. That's what Charles was teaching me.

He asked if I would join him on a trip to London early in the new year. "Yes," I heard myself say.

Yes, yes, yes.

•

That Christmas Eve, after Max and Alex had gone to bed, I sat down at my kitchen table and pulled out a sheet of my family stationery, cardstock in cream. In the top right corner, it reads *The Seagers,* embossed in blue cursive. I found a pen and wrote the date in the top left: *12/24/2013.* I addressed it to Dr. D., Mike's cancer doctor. Although I was nearly through with my meltdowns, holidays were still hard. There was still something triggering about seeing, even imagining, the unburnished joy of others. I took a moment to gather my thoughts and began to write.

While you celebrate Xmas with your happy family, the boys and I mark our third without Mike.

And then I vented. I vented about what he had done to Mike—not his failure to save him, but the damage to him that he had done in his attempts. *Three years ago you insisted on a third type of chemo, one with a known chance of success of 0.000000000%. I was shocked to later learn you only cared about covering your own ass, if after death the family were to complain you hadn't done everything you could.* I had wanted to end my time with Mike the way we had started our time together. I had wanted Mike to die feeling strong. *All I wanted was one chemo-free month before cancer took over—just one more great month with Mike.* I wanted him to feel that he had beaten cancer, not because he had survived it, but because it had never dictated how he might live. That's not what happened. *You ruined Mike and my precious time left together.* Part of me was still angry about it. Part of me might always be angry about it, but I wanted to try not to be angry anymore. I dug my pen into that sheet of paper and told the doctor that I was in charge now. I was in control. Not him. Not cancer. Not the universe. Me.

You owe me an apology, and I am still waiting.

I looked at that piece of paper on the table for a long time. I had rid myself of the poison inside me. All the pain, all the hurt, all the regret, every trace of bitterness and rage, I had spilled onto that single page.

I never sent the letter.

CHAPTER 20

Final Report

Charles and I met in the polished halls of Heathrow on the first day of 2014. I hadn't even thought to cry at midnight. I had arrived first and waited for him. We found each other at Arrivals. There was no turning back. We went out for dinner that night. I asked him, the way two kids new to love might ask each other: "Are we boyfriend and girlfriend?" I needed to be clear about what we were and what we were going to be.

"Yes," Charles said.

"Are you sure?"

"Yes."

We had booked separate rooms again, because—well, I don't really know why. Habit. Nerves. A sense of propriety. We were both anxious and hazy from more than the jet lag. Early in our trip, I caught Charles looking at me. "You're really very beautiful," he said, and he raved about my figure, too. He said it as though he had never realized it before, my body a surprise discovery. We had fallen in love mostly in two dimensions. Only now did it occur to him that I wasn't an apparition. I was a real, physical object. We might also touch each other, and it might be nice to touch.

We embarked on a grand tour of London together. Charles took me to the Royal Observatory in Greenwich, split in half by the Prime Meridian Line. We lingered in the Clock Room, looking at the evolution of timekeeping and the charting of stars. It was a little on the nose, the two of us standing together, unsure of how or where our future together might finally begin, watching countless ticking clocks.

London was very London that first week of January: cold, damp, thick with fog. Charles felt like a fire next to me. Despite the chaos of a great world capital swirling around us, my eyes were nearly always on him. But there remained an uncertainty between us, a tentativeness born of disbelief.

One night we went out for dinner, one of those extraordinary experiences—the food, the wine, the ambience—that you know you will remember as long as you live. Afterward, Charles led us on a walk through the empty streets of late-night London. There was only the sound of our boots on the cobblestones. We turned a corner and came out of the mist. I still have no idea how we came to be standing in front of Buckingham Palace.

"You are my princess," Charles said.

We kissed.

It was worth the wait.

We flew home without making plans for what we might do next. Charles called me and left the sweetest voicemail. It was also endearingly awkward: "We should have made plans," he said. "Let's make some plans. I . . . Well, I was wondering . . . If you'd like to come to Toronto. Or maybe I could come visit you?"

We did both. First I went to Toronto to see him and meet his parents. Next I invited him to come see me in Concord. I wanted him to meet the boys. Max had turned ten; Alex was eight. Charles knew that we were a package deal. He wanted to make the right kind of impression on them. I have a photograph of him with the boys from that night; I can still feel the weight of that meeting in every pixel, the expectation and desire of it. Even Alex could sense it. He pulled out an enormous book of M. C. Escher illustrations that he thought Charles might like, and opened it across both of their laps.

I had used some of my mountain of air miles to book us a February night at the Boston Harbor Hotel, a luxurious stay in the city while Jessica looked after the boys at home. We were out at dinner when Charles began talking about the future in a way that puzzled me. His thinking seemed long-term and specific, the way I might talk about the stages of future space exploration: *this,* then *this,* then *this.* He talked about how the boys and I could come up to Tiny that summer to stay at his cottage, roasting hot dogs around the fire. Maybe he could show them the Green Flash over the bay. He had made other plans for us in the fall, and the following winter, and the spring that would come after the spring that hadn't yet arrived.

"Charles, we've just started dating. People don't normally talk like this. Not so soon. Not so early."

Charles looked down at the table. I worried that I had upset him the way I so often upset people, saying things that were better left unsaid. He was, in fact, finding the courage to do the bravest thing he had ever done.

He looked up: "I've been thinking about this for a long time," he said. "Sara, will you marry me?"

That's how Charles said it. There was no preamble, no foot- or endnotes, none of his usual jokes, no hesitation. He was direct and clear and certain. He had no time to waste.

Neither did I.

"Yes."

The next day I got home and told Max and Alex the happy news that I'd been struck by lightning: "Charles proposed to me last night, and I said yes!" It sounded strange coming out of my mouth. To me it sounded too good to be true. That's not how it sounded to Alex. "*What?*" he said. He said it with a surprising anger, as though I'd told him something that he refused to believe. "You should have consulted us first!" I had been so careful in how I handled Mike's death with them, in deciding what they had needed to know and when, and what they could go without knowing. Now I felt as though I had threatened all of that delicacy in an instant of abandon. That was the first time love had made me careless, not careful.

The next evening, I got home late from work. Veronica had provided that night's cover. I went upstairs to check on the boys. I could hear Alex whimpering in his bed. He was sharing a room with a sleeping Max, and Alex was being considerate in his sorrow. His attempts to stifle the sounds he was making left me only more heartbroken.

I sat on his bed and spoke in a whisper. "What's wrong, sweetheart?"

"My whole life is going to change," Alex said. "I don't want it to change. I like my life." He had been doing some very adult math. He was worried that we wouldn't be friends with the Widows anymore; we wouldn't go on trips anymore; Jessica and Mary and Vlada and the rest of our gang wouldn't live and travel with us anymore. Our friends had filled the empty spaces. Adding Charles meant that there wouldn't be the same kind of room.

I told Alex that our lives would change, but they wouldn't change quickly, and they wouldn't change for the worse. Nobody he loved was leaving. Somebody he would come to love was arriving. Life was good. Life with Charles would be even better.

I didn't believe everything I was saying. I felt as though I'd made a huge mistake. If the boys didn't see in Charles what I saw in him, I would be in an impossible place. There was no way for me to choose between them. I had to hope that everything would correct itself, that the planets would align. That's where I put my faith: in Charles, and in my boys, and in each of their capacities to love and be loved.

"Just you wait and see," I said.

•

We published our interim report on the Starshade that April. We had done good work. The Starshade wasn't seen as an impossibility anymore—even some of the harshest former skeptics now believed. NASA officials who had expressed an almost physical discomfort around me during our early meetings now beamed when they saw me. I wasn't a propagator of lunacy; I was an ambassador from a miraculous future.

We still had a long way to go. I had made some of my six mandated trips to work on the science and technology of our beautiful machine. The engineers at the Jet Propulsion Laboratory came up with our final design: a flower with a circular center about twelve meters across, with petals that were seven meters from base to tip. The Starshade's hardware needed to be manufactured to tolerances measured in hundreds of microns, which meant that a large part of the challenge lay in our physically constructing it. I felt as though we could manage. Model petals made of aluminum and composite had been built, and they seemed to work well. It was a thrill for me to watch them release themselves as though from a bud. They were something special to reach out and touch. We folded them back up and unfurled them again, eventually deciding to add a system of mechanical shims to find the perfect balance between fragility and strength. Once the Starshade was in space, we'd have one chance to get things right. Fingers crossed wouldn't cut it. We needed to know.

There were still problems. One of them was the brightness of our own sun. Its light, too, would bathe the Starshade, and, at certain angles, what we were calling "solar glint" would reflect off the petal edges, interfering with our images. Another massive design challenge would be figuring out how to keep the Starshade flying in formation with its distant space telescope. They would be orbiting tens of thousands of kilometers apart and yet would need to line up with mathematical precision, our shield slipping between the telescope and its target star. It would have to be able to move, too, and to repeat that ballet. We had reduced the number of star systems we might examine to perhaps two dozen in its lifetime, and maybe in mine, because moving requires fuel, and fuel means mass. But the Starshade would still demand some of the most complex choreography in human history.

I liked our chances. Despite our initial friction, our committee had found an undeniable chemistry and purposeful rhythm. I could almost see the connections building between us.

My life at home, through its own kind of alchemy, was feeling more and more whole. Charles was coming down to Concord every other weekend. We might have known what we wanted from each other quickly, but we took our time in completing the transaction. Thinking back, those probing weeks and months make me think about how we might approach aliens after we make first contact. We were cautious when we first sent astronauts to the moon. When they came back to Earth, we put them in quarantine, on the deck of a ship in the middle of the ocean, in case something sinister had been hiding in the dust. After we find proof of another life in the universe, I imagine we will take our time to decide whether it's a life that we want to know.

That's how it was with Charles and me. We knew we loved each other. Our connection was obvious. But we were careful in the finishing of our fit. Most important, I wanted to know that Max and Alex were as happy as I was.

Because I was really happy. Charles was smart and curious and funny. Even when I was feeling stressed and overdrawn, he always found a way to make me laugh. He was supportive about my work without being intrusive about it. He never asked me a question that started with "Why?" He knew why I cared about the stars, and what that love of mine might mean. He knew the feeling that comes with looking through a telescope, the bigness and smallness, the knowledge and the mystery. He knew how demanding our shared love could be, too. He didn't ask questions, because he knew that I would never have all the answers. He knew how endless the universe is.

Charles was also helpful in practical ways, good with his hands, familiar with tools. In an hour he would cross off a job that had been on my to-do list for years. He spun delicious meals out of the same kitchen in which I'd struggled. "I want to make you the happiest woman in the multiverse," he told me, and I wanted to let him try. It took me a while to accept that he might. In the beginning I wondered whether the men at Rocky's Ace Hardware might miss me. I had enjoyed my budding self-reliance, the satisfaction of a small job well done. I didn't necessarily enjoy housework, but I had liked the feeling that the work had given me. I think I'd found in those jobs a kind of security: I could survive anything; I could survive being alone. But every time Charles put dinner in front of me and smiled at my smile, every time I came home and the fridge was full or the snow had been shoveled or the battery in a smoke alarm had been replaced, I learned to accept a little more of what Charles wanted to give me. He wanted to give me a different kind of peace. There is more than one way to feel complete.

Max and Alex eventually fell in love with Charles nearly as deeply as I did. They saw how much happier I was with him around, which meant they were happy when he was around, too. After his second or third weekend with us, Alex took me aside: "How soon can Charles move in?" Not long after that, Alex asked me what he did for a living, and I told Alex about his family business. He took careful note. Apparently, kids at school sometimes talked about their parents, and Alex had learned to hate the silence that followed questions about his father. By that spring, Alex couldn't help pressing: "How soon can you and Charles get married?" His birthday sleepover was coming up. He wanted to introduce Charles as his dad.

Home and away, I was in the middle of two jobs that were something like the same, each with its own pressing deadline: With an improbable collection of parts, I was seeking to build machines of grace. I knew all of our needs—my Starshade colleagues', my family's. Now I had to help supply the most elegant solutions that I could find to meet them. At the center of both of my assignments were unlikely unions. The only difference was that one of them would come together to destroy light. The object of the other was to deliver it.

•

That December, the Jet Propulsion Laboratory made me a fantastical interplanetary travel poster. I thought it was magical. Later a whole series was made. They look vintage, like old

posters for extinct airlines and train trips across the "mysterious Orient," but we've since set different sights, and their des-
tinations are fully forward-looking. They were such a hit that
the lab made them available for download; the site soon
crashed from the demand.

Some of the posters advertise visits to planets within our
solar system. There's one for Venus, with an observatory
floating above the endless clouds, and one for Jupiter, where
hot-air balloon rides take you closer to the mighty auroras.
There's one for Ceres, the Queen of the Asteroid Belt—and
the last stop for water before Jupiter—and one for Europa,
where maybe life waits for us under the ice. (We might never
be able to see more distant moons, but if you start counting
moons as possible homes, the odds tilt even more in favor of
other life.) Importantly, I think, there's also a poster for Earth,
featuring a pair of astronauts sitting on a log, looking out
over lakes and mountains and trees. Imagine an alien seeing
Earth for the first time.

But my favorite posters fall under the guise of the Exoplanet Travel Bureau. They are the ones that trumpet visits to
worlds that we are just beginning to understand. There is one
for Kepler-16b and its magical twin suns, "where your shadow
always has company." A future explorer stands between rock
faces, his pair of shadows stretching out behind him. There is
one for Kepler-186f, with a white picket fence strung across a
scarlet landscape, "where the grass is always redder on the
other side." Trappist-1e, part of a tight knot of seven rocky
exoplanets, is imagined as a stepping-stone, an interplanetary
way station. HD 40307 g is portrayed as a place for skydivers
to try their luck against the gravity of a super-Earth. There's
even a poster for PSO J318.5-22, the rogue planet, trapped in
its permanent midnight, swept by its molten-iron storms. It

depicts a glamorous couple, dressed for a gala, posed arm in arm. They've come to the place "where the nightlife never ends."

I had the exoplanet posters printed and mounted in frames, and I hung them in the halls outside my office. I loved walking by them every day on my way in and out of work. I do today. I look at them and see Charles and me.

•

One Wednesday afternoon, Chris came to see me at MIT. We did what we usually did during her visits: We talked about our kids and work and summer plans as I tried on dress after fashionable dress. I usually bought one or two, but she never forgot to tell me that I still wore too much black and needed to burn my hiking boots. I protested that black goes with everything, and I hardly ever wore my hiking boots anymore. I could hear her wry remonstrations whenever I even thought of putting them on.

Afterward, walking her back to her van—it's black, by the way—she grabbed me by the arm and looked at me with far more serious eyes than she usually did.

"Does finding love again mean the pain is gone?" she asked.

I didn't know how to answer her with words. I shook my head.

That night, I had my usual dream about Mike: He showed up back in my life after a long time away. This time, he'd been in a coma. The difference now was that I had found Charles. I told Mike that I was engaged to a man I loved. Mike told me that he understood—he was calm, even reasonable—but that I had to end it with Charles. We had to go back to the way things were.

I woke up with a start. I realized that I had to say goodbye to Mike, with the finality of a door slammed shut, before Charles and I got married. I would force myself to imagine Mike's coming back to me, as he did in that dream, and I would do what I had to do: I would tell him that I chose Charles. I had to break up with Mike in my imagination again and again. I broke up with him when I was staring out the windows on the train. I broke up with him in my office. I broke up with him when I was the last one up at night. Every time, I said the same four words: "Mike, I choose Charles."

I kept my imaginary dialogue private. I didn't tell even the Widows about it, because I knew that some of them would disagree with what I was doing, strongly. Some of them believed that you stay married to your dead husband forever, no matter if someone new comes along. But I knew that I couldn't commit to Charles until I was no longer committed to Mike. When someone you love dies, you don't leave them. They leave you. Your love gets stuck, lost in translation between this world and the next: You're constantly giving your heart to someone who isn't there to receive it, and at the expense of someone who is.

A few months later, I had another dream about Mike. He was coming to me less and less. This time he was in a wheelchair, paralyzed from the waist down. He'd been in an accident, and it had taken him years to recover. He looked good, though. His hair was red again, not chemo-gray. I think I heard him before I saw him, digging through the house looking for some piece of gear, but he seemed more than distracted when I found him. He was hiding something from me.

Then I saw her. A new woman. Younger, with red hair like his. She was pretty but not too pretty. She was helping him with a contraption that would allow him to still use his kayak, which made sense to me: You don't use your legs in the boat.

In my waking life, I had continued to slowly get rid of our boats. I'd donated my Dagger Rival and the family Swift Yukon to the MIT Outing Club; I wanted to keep those two within reach, just in case I ever felt the urge to paddle again. Most of the rest of them were gone. But in my dream, I could see why Mike still wanted to be on the water. In the boat, Mike could be who he had always been. I felt the smallest blush of surprise—*Oh, okay, you've moved on*—but mostly I was happy for him. Nobody was hurt or hurting. Nobody was alone anymore.

I woke up. That was it. That was the last dream. I never saw Mike again.

•

The Widows celebrated one final Father's Day in the summer of 2014. Chris hosted us in Lexington. Not everyone came, and there were hardly any kids. Max and one of the other sons ganged up on Alex, and he ended up in tears. I brought Charles to meet everyone, but we were in our own rush, headed for the airport for his trip back to Toronto. It was all strained. That marked the end of our formal gatherings. If I see the Widows anymore, it's mostly by accident, at the park or the grocery store. There was one month not long ago when I ran into all but one of them, an unexpected burst of electricity through dormant circuits. I'm always happy when I do see them, but there's something a little wistful about our chance encounters. They feel like they should be more purposeful.

Melissa is the only Widow I still see regularly, nearly every week. She's still my best friend, even though I know she has other best friends. I see Chris, too, but not as much. I run into Melissa on the train into Boston; we meet for morning dog walks; once in a while we get our nails done. She tells me about her life and I press her for details about her latest men. I tell her about my latest planets. She still solves my problems for me, but they are so much smaller than they once were. I smile all the time I am with her.

The rest of us have grown apart. I suspect it's because our lives became even more divergent than they already were. We all reached different versions of normal; time brought our differences closer to the surface than our similarities. We're all busy with whatever we did before we became widows. Melissa went back to work with Fidelity in Boston's financial district; Chris has built a thriving interior decorating business. Sometimes, someone tries to set up a dinner. Sometimes I reach out. A few of us might show up, but the full membership of the Widows never seems to manage it. If you had pulled me aside the last time we were all in the same room and told me that we were spending our last hours together, I would have assumed that something else terrible was about to happen. Which one of us was going to die? But our end wasn't like that. There was no trauma, no cataclysm. We called a little less. We emailed a little less. Nobody brought anybody plants anymore. We all had enough plants. Some of us found new men; some of us didn't; some of us never tried. Bit by bit, we drifted apart. Our initial connection had been born of our losses. Our subsequent victories didn't seem to have the same effect.

One day Charles made a suggestion. Or maybe it was a request. He didn't like that I still referred to the Widows of Concord as "the Widows."

"You should just call them your friends," he said. That's what I called the other people who had been there for me. I called my students my friends. I called my helpers my friends. It would be evidence of my recovery, using the nomenclature of survival. It was also a matter of accuracy. But I still think of myself and my friends as widows.

Maybe one day the Widows—my friends—will all get together again. I know in my heart that they are still out there for me, and I will always be here for them. At first I wondered whether they were like most of the other relationships in my life up to that point: utilitarian and transactional, means to an end. The Widows were there to help me share my grief and my pain. My postdocs and students, like Brice and Vlada and Mary, were there to make me feel hopeful. Jessica, Veronica, Diana, Christine—I had found them to lighten my load. That's what they were at first. They each had a specific purpose. But all of them became something more to me, as I became something more to them. Our relationships became less about need and more about want: I wanted to spend time with them; I wanted to help them; I wanted to listen to them. If the Widows were still a kind of instrument, they were more beautiful and precise than I ever imagined they might seem to me, a shimmering collection of sextants and compasses to help me navigate my way through not only Mike's death, but the rest of my life. We've moved on from each other in a lot of ways. But sometimes I still look at them, waiting in their places in my mind, and I think about how good they feel to have around me, the warmth of their shine, the comfort of their weight in my hands.

·

In March 2015, we issued our final report for the Starshade. It is 192 pages of charts and spreadsheets and illustrations that can be boiled down to a single line: *We know how to build it.* Together, we had solved or had a plan to solve each of its de-

velopmental hurdles, and it's become an official NASA technology project because of our team's work. That means it's a
real investment with real dollars. It could be a real thing. The
answer is a flower with a huge center and as many as twenty-
eight pointed petals: less a sunflower than a child's drawing of
the sun. I'm certain that it's the best way for us to take direct
images of exoplanets. Not just any exoplanet, orbiting some
angry red dwarf—another Earth, orbiting another sun. I think
it's the most beautiful spacecraft in the universe.

The Starshade would help a space telescope see differently
than Hubble, differently than Kepler. Think of what they've
allowed us to see. Today, right now if we want, we can do
more. What is maybe my favorite illustration comes late in the
report, a dramatic before-and-after comparison, demonstrat-
ing the difference that the Starshade would make to a space
telescope's field of view. In the before, the telescope is blinded
by the light of a single powerful star. In the after, there's a
near-perfect darkness. It's as though the star is an all-
consuming fire, and we've learned how to blow it out like a
candle.

We also came in, for the time being at least, at $630 mil-
lion, well under our billion-dollar budget. That's if we rendezvous the Starshade with a space telescope already on the
books—perhaps the Wide Field Infrared Survey Telescope
(WFIRST), scheduled for launch in the mid-2020s. We know
that we're still asking for a lot of money. A lot of other things
can be bought with $630 million. The Air Force has plans to
replace its current fleet of Stealth bombers with a hundred
new B-21s. In 2017, the Congressional Budget Office esti-
mated the cost of the program to be $97 billion. In other
words, we could build one fewer B-21 and build the Starshade
instead, with a few hundred million dollars left over. We just

need to decide as a country, and as a species, what sort of future we want. What matters to us and what doesn't? What do we want to accomplish? How do we want to be remembered?

Our report had to be temperate and academic in its language. "The Exo-S mission study serves as a proof-of-concept that a low-risk, cost-driven, $1B 'Probe-class' mission leveraging proven technologies is capable of ground-breaking exoplanet science," we wrote. "Exo-S would be a major step toward directly revealing the planetary systems of nearby stars, with luck finding a planet as small as Earth . . . It is our sincere hope that the results of this study will prove useful to the design of a future imaging mission for the study of habitable exoplanets."

I am less temperate in real life, one of the lasting virtues of my grieving years. By the end of the Starshade study I'd returned to the team working on TESS, the wide-scanning space telescope being developed at MIT. What had seemed too much to bear after Mike had died now seemed relatively easy, at least in comparison to the Starshade's many challenges. The difference between the two was that TESS was a physical reality; it would launch in April 2018. (I was its Deputy Science Director from 2016 to 2020.) Seeing it take physical shape in the lab only made my desire to build the Starshade more intense. Why was one possible and the other not? Why couldn't it be made real, too?

I borrowed a small-scale version of our Starshade, used for testing in the desert, and a full-size model of one of its petals. They fit into a pair of black travel cases. Those cases are now as battered as the doors I have beaten down trying to get the Starshade made. I have taken those models into classrooms and airports and lecture halls. I have tried to show people how we can explore our universe more completely than we might ever have imagined. We could prove that we're not alone. Children almost always understand what I'm trying to say. It's the adults who sometimes don't. Adults have been given too many reasons not to believe; that's why they so often say no. Children have greater faith in us. That's why they always say yes.

•

I married Charles a few weeks after we published our final report. When we went to get our marriage license, we asked the town clerk whether she could marry us. She could, but the ceremony would have to take place outside the office, after hours. She told us about a park across the street, and we headed out to take a look. It was perfect. There is a little bridge that crosses the brook that meanders through Concord. That's where Charles and I would cross to our new lives together.

Of course, Melissa helped me pick out my wedding dress. She also helped me hire her friend to take photographs: Gigi, the same photographer who had shot Chelsea Clinton's wedding and our online dating photos. I wanted evidence of our love that I could hold in my hands. The day of our wedding was cold but bright and clear. It was the right kind of spring day, one of those on which the world seems about to burst with life. I had my makeup done, put on my dress, and headed to the park. The only people there were the clerk, Gigi, Charles, and me. That's what we wanted. We said simple vows to each other, honest expressions of our love and commitment, sometimes pausing to laugh, sometimes to cry. I like to think that Charles and I each rescued the other. During the ceremony, my brain kept asking itself the same questions. *What were the odds? How did I find Charles? How did he find me?* I felt like the luckiest person on Earth. I still do. After feeling cursed for so long, it is blessing enough to feel blessed.

After the ceremony, Melissa met us at a restaurant a few blocks away in town, and Max and Alex joined us. Veronica came along, too. Melissa had organized a champagne toast. She told us that true love is nearly impossible to find. Another Earth should be easier. I knew how lucky I was to find her, too, that long-ago day on the hill.

The boys started climbing all over Charles, telling him how handsome he looked in his suit. They didn't pay any attention to me, in my pretty white dress and makeup. I mean, Charles looked dashing—*but hello, blushing bride over here!*

Alex didn't take long to ask the question that had been on his mind for a year: "Can we call you Dad now?" The grown-ups in the room felt the answer catch in our throats. Melissa was in tears.

"Yes," Charles said finally. "Absolutely."

Then Alex asked if he could have some champagne, and Charles poured him a little.

"My head feels funny," Alex said.

"Charles!" I said. "Your first official day as a dad and you got your kid drunk."

Alex knew a soft touch when he saw one. After his champagne, and a root beer, he asked if he could have one more thing.

"Sure," Charles said.

"You know that ice cream place—"

That's when I stepped in. "I think it's time to go back home, boys."

Charles and I spent our first night together as husband and wife at the inn that's in the middle of Concord. We woke up the next day and went to the town hall to pick up our marriage certificate. We walked home together with it in our hands, our proof of another life.

·

Not long afterward, we filed more paperwork, this time for Charles to adopt Max and Alex. We wanted to make his fatherhood official. Freya, the lawyer I had found when all I wanted was a haircut, prepared our application for us. A widow and a lawyer, twice useful, she never had a chance to make her escape; I was never going to let her go.

We had to appear in family court in Cambridge. Max and Alex wore their first suits—a little ill-fitting, but they were such handsome young men. Charles was a handsome, slightly older man. Even in the winter he kept his color, and I still loved the way his crisp white collar looked against the strong line of his jaw. A big storm had been forecast for that day, snow up to the windowsills. We didn't want to risk missing our morning appointment, and we stayed the night in a hotel in the city. We woke up to a pretty dusting of snow that soon cleared, leaving us under the watch of a bright winter sky. The storm never arrived.

Freya met us at court, along with her assistant lawyer. The judge and the bailiff were the only other witnesses to our important moment. Family court is normally a hard place. Most families go there because they are falling apart, and falling apart in a way that causes them to need strangers to help settle their differences. There is so much sadness in those halls. The judge had a reputation for being tough, but I think she was almost relieved to see us. We stood before her so that we could come together.

The judge read our adoption statements, first for Max, then for Alex. She had them each bang her gavel to make the other's adoption official. Now Charles really was a father, and the boys again had a dad. I couldn't believe it. There had been so many black days and sleepless nights when I never could have imagined something like this scene, at least not with me in it. I had struggled for years with the most challenging math, and now here it was, solving itself in front of me. We made four.

We walked down the courthouse steps together. A family. We checked out of our hotel and got into our car. Charles asked the boys if they had buckled themselves in. "Yes, Dad." He started up the engine and turned the wheel toward home. The sun was still winter-high and shining. The sky was cloudless. There wasn't a shadow in the world.

We had asked the boys if they wanted a big party with their friends to celebrate, or a quieter gathering at home, just the four of us, sitting around the table in four chairs, eating some cake.

They chose cake.

·

When Charles moved in, he asserted his presence in quiet ways, in order to feel more at home in a house that didn't always seem like his. The wires that I had left taped and dangling from the ceiling in the front hall bothered him. They looked bad and possibly dangerous. He asked me why there wasn't a light there, and I told him the story of the boys and their duels, and how I had worried about discovering a hurt child or a broken light, and how I had taken down the light all by myself, and how I had felt reduced and gigantic at the same time. There'd been a hole in the ceiling ever since. I didn't know how to fill it.

Charles was always good at taking my hints. One day when

I was at work, he seized his opening: He got out the ladder, carved out the plaster, installed a proper junction box, ran the wires through it, and hooked up a new fixture. Flush-mounted this time, too high for even our growing boys to reach.

Charles made sure that the light was on when I came home that night, filling our front hall with its orange glow. It spilled out of our windows and onto our front steps, and when I came up our walk, I felt its warmth. I stood outside and looked through our windows for a long time before I walked up the steps and opened the door to the sounds of happy boys and the smell of supper.

Sometimes you need darkness to see. Sometimes you need light.

CHAPTER 21

The Search Continues

In August 2017, after years of work and hope and effort, SpaceX prepared to launch a Falcon 9 rocket from the coast of Florida into space. The rocket didn't have a crew, but ASTERIA was on board.

It had been a difficult journey. The camera had made its way from my imagination to our design-and-build class, through drawings and prototypes and an old missile site in New Mexico, to under the seat on the plane with the protective Mary Knapp. Then we'd run out of money at MIT, and Draper Laboratory had liked the technology better for other things. The Jet Propulsion Laboratory, which had always been interested in the possibilities of CubeSats, ASTERIA especially, picked up where MIT and Draper left off. Three MIT graduates there would play leading roles on the project; they took their work seriously, having seen firsthand how much it mattered. Their passion and expertise made sure that ASTERIA would become everything it could be, that it was built right and lovingly placed, at last, into the hold of a rocket, groaning on the launchpad on a beautiful late-summer day. The rocket would slice into the sky and rendezvous with the International Space Station. The astronauts there would set free our little satellite later in the fall. From a little whisper in my dreams to space: I couldn't believe that we were nearing the end of such a long reckoning. The last time I'd watched a rocket launch, it was when Kepler went into space. We'd found so many new worlds since.

Charles and I had planned on going to the ASTERIA launch, but it was delayed by a few days, just long enough for our travel and childcare plans to fall through. On the day of the launch, I took the train into Cambridge instead, walked to the Green Building, and took the elevator to my floor. I walked past the travel posters for distant worlds into my office and shut the door. I sat alone. I opened up my laptop and called up the online video stream. The launch was a big deal to a lot of people for a lot of reasons; all over the world, eyes were trained on that rocket, still waiting on the pad.

My curtains were opened, and every now and then I looked up from the cloudless Florida footage on my screen and out my windows, at my crystalline view of downtown Boston. There were clear skies everywhere I looked. The launch was scheduled for 12:31 P.M.

I spent maybe thirty minutes in the quiet, writing thank-you emails to other members of the ASTERIA team. At the last second I decided not to send them. I know that superstition is unscientific. I understand that it doesn't matter to the universe if a baseball player is wearing his lucky underwear—whether he gets a hit is mostly up to the pitcher and to him. But rockets are delicate, ill-tempered machines. Before the Russians launch their rockets from the steppes of Kazakhstan into orbit, they summon an Orthodox priest to throw holy water at the boosters, his beard and cloak and the holy water carried sideways by the wind. I wasn't going that far, but I wasn't going to send a couple of emails until we were safely weightless.

I was surprised by how nervous I was, watching the countdown clock tick down to launch. Finally it reached zero and I leaned so close to my screen that someone watching might have thought I was trying to crawl through it. In a way, I was.

The engines ignited with a great big ball of pure fire. The launch tower fell away, and the rocket began to ease its way off the pad. Rockets are almost alarmingly slow in the first few seconds of launch. They seem more like a container ship trying to grunt its way out of port than what they are. But after a few seconds, that Falcon 9 really started moving. It flew straight up. Its ascent began to curve a little, and it pushed its shining shoulders toward its future orbit. The onboard cameras recorded its arching flight. In what seemed like seconds, the sky around it went from blue to purple to black. The rocket had broken through into space. The boosters were jettisoned, and the remainder of the rocket continued its climb into the deepest possible night, the Earth blue and alight behind it, an impossible blackness ahead. It would take a little while for it to catch up with the space station, which was racing its own way through orbit at 17,000 miles an hour, about five miles every second. But the rocket, and our satellite, were well on their way.

Everything brave has to start somewhere, I thought.

●

Not everything brave has to end.

Late in 2016, after Charles and I were married, before our satellite had launched, *The New York Times Magazine* published a long profile of me. I was flattered by the attention. I was also a little worried about the story. It can still be hard for me to be open about myself, and people I know talked to me after the story came out and said they were surprised at how much I had shared about my life. "It's very personal, Sara," one colleague said. He was right. It was. A good friend was angry: "The writer went too far." I thought all of my secrets were out.

But I also learned something about myself in that story. After he'd seen the article, Bob Williams emailed me. His wife, an experienced autism specialist, had read between the story's lines. She could spot an autistic person from far away. The way they walk, the way they move their hands. The way they're often alone. Up close, the signs become unmistakable. Autism is in the eyes, unblinking. In the voice, monotone. In obsessions with machines and how things work. In an attention that can never be shaken.

She thinks you have autism, Bob wrote. *I've never met anyone who can focus like you.* I told him that his wife was wrong, it couldn't be. I was too old not to know such a basic fact about myself. No one had ever suggested that I was anything other than quirky as a child, strange as an adult. Back and forth, back and forth. Finally I went to see a doctor who specializes in diagnosing psychiatric disorders, and she confirmed what Bob and his wife already knew. I saw myself clearly for the first time.

I can't begin to explain the *truth* of that revelation. I felt it land, as though I'd been struck by something, a physical impact. So much of my life suddenly made sense. I thought back to my lonely childhood. To my thirst for wide-open spaces and the mysteries they might hold. To my solitary quest for connection, and the way people look at me when I try to talk to them. To my love of logic and the stars. To my refusal to believe that we're alone, that there is no one else out there like us.

Of course. *Of course.*

Sometimes we think we know what we're going to find, and where we're going to find it, and we don't. Sometimes, as with ASTERIA and another Earth, we know what we want to find, and we think we have the right way to find it, but we still don't know that we will. Sometimes, as with Charles and me, we find what we're looking for most in the world, and we might never understand how or why we did.

And sometimes, if we're really lucky, maybe only once or twice in our lifetimes, we find something we didn't even know we needed.

I've started to wonder whether that's the best kind of science: the revelation that's equal parts unexpected and essential, the accidental necessity. Isn't that the most thrilling form of exploration? It's more satisfying than a question without an answer. And it's more profound than an answer without consequence. There is no greater breakthrough than the answer to a question that we never thought to ask.

Is there other life in the universe? That's the question that I've always thought I needed to answer. Maybe I've been wrong all this time. Maybe trying to see the smallest lights in the universe isn't about who, exactly, we'll encounter there. Maybe it doesn't matter what aliens really look like, what form their version of life takes. Maybe our search shouldn't be about them. Maybe it never was.

Do I believe in other life in the universe?

Yes, I believe.

The better question: What does our search for it say about us? It says we're curious. It says we're hopeful. It says we're capable of wonder and of wonderful things.

I don't think it's an accident that there's a mirror at the heart of every large telescope. If we want to find another Earth, that means we want to find another us. We think we're worth knowing. We want to be a light in somebody else's sky. And so long as we keep looking for each other, we will never be alone.

Acknowledgments

The story of my life first became public when Chris Jones wrote about me with consideration and heart in *The New York Times Magazine*. I'd like to thank Chris for his help with this book.

Chris and I were connected, in ways that are too cosmic even for me to explain, by Matt Reeves, Rafi Crohn, and Adam Kassan at 6th & Idaho. Mollie Glick at CAA also gave me crucial assistance in navigating this new universe. I am indebted to them for their kind interventions.

At Crown, my appreciation to Rachel Klayman for seeing the potential in the smallest of lights; Gillian Blake, Meghan Houser, and Lawrence Krauser for their direction and fine-eyed edits; Mark Birkey for his careful production-editing work; Elena Giavaldi for her beautiful cover design; and Gwyneth Stansfield and Rachel Aldrich for their help in spreading the word.

Because Beth and Will appear too briefly in the book given their importance in my life, I want to give them my deepest thanks here for their love and generosity. Their Christmas tree farm was a safe place for my family when we needed it most.

I have been blessed with other havens. I am beyond grateful to my colleagues, postdocs, and students at MIT, without whom I might not have come close to healing during the difficult time spanned by this book, as well as my father figures and friends from among our generous and supportive alumni. My endless admiration, too, for the scientists and engineers at the Jet Propulsion Laboratory and Northrop Grumman Corporation, and elsewhere across the country, who are dedicated to the Starshade and other space-based direct imaging missions. One day we will find what we most want to see.

I must also thank, very much, the Widows of Concord for your unique understanding, unwavering support, and impeccable fashion advice. You helped me when I needed more help than I ever feared I might, and I will never take your friendship for granted.

To Jessica, Veronica, Diana, and Christine: You will always be family.

My love and gratitude to my wonderful boys, Max and Alex, for your patience and the joy you bring to my life. I am so proud of who you are.

And to you, Charles Darrow: Thank you for introducing yourself to me at that salad bar in Thunder Bay, and thank you for rescuing me every day since. I will love you forever.

THE SMALLEST LIGHTS
IN THE UNIVERSE

Sara Seager

A BOOK CLUB GUIDE

Questions and Topics
for Discussion

1. The author opens the book by describing rogue planets; she uses them as a metaphor for her children, who she says have gone "halfway to rogue" following the death of their father. What else in her life appears "rogue"? Who or what in your life could be described as a "rogue planet," with no star to orbit?

2. Throughout the book, the author talks about the power of belief and positive thought. Do you feel that belief is a type of magic? Why or why not?

3. The author is an extremely successful woman in a field dominated by men. Was there a point in the book when you thought this circumstance was especially affecting her? Do you think the fact that she's a woman has had an impact on her career trajectory, for better or for worse? Why?

4. Is there anything in your life that you've pursued with blind faith despite opposition, in the way that the author is driven to find exoplanets in the face of backlash from her

scientific community? What kept the author moving toward her goal? What keeps you moving toward your goal?

5. Later in her life, the author discovers something about herself that she had never considered before—she realized it only after she was featured in a major publication and a friend pointed out certain aspects of her personality that came through on the page. How might you have reacted to a surprise like this? Have you ever realized something about yourself only after seeing yourself from another person's perspective?

6. The author relied on a dark sense of humor to cope when her husband was first diagnosed and throughout his illness. What do you make of this? Why is this her instinct? Does this form of humor appeal to you, or not?

7. When her husband passed away at home, unhindered by tubes and machines, the author says she felt she was able to help "build something beautiful." Do you agree that death can be beautiful? Why or why not?

8. What do you make of the use of metaphors throughout the book such as dark and light or the sun and stars? Was there a particular metaphor that was the most powerful to you?

9. The Widows of Concord become a supportive community for the author after her loss. Why do you think the author initially resisted their friendship? What did she ultimately gain from those relationships?

10. In her recurring dreams of her husband following his death, the author sees him return to her after long absences: he has been in a coma, missing, on long trips, and so on. What do you think is the meaning of this recurring dream?

11. Do you feel that the scene with the Green Flash is a moment of rebirth or closure for the author? Is it—or *can* it be—both?

12. The author has focused her life's work on detecting life on other planets, only to find herself searching for new life after death. How are these pursuits related? How are they dissimilar?

A Q&A with Sara Seager

Random House Book Club: Sara, you won a MacArthur Genius fellowship in 2013. The citation said you earned the grant for "adapting fundamental maxims of existing planetary science to create a comprehensive theoretical framework for determining the characteristics of planets beyond our solar system." Could you talk about your work in slightly more accessible terms?

Sara Seager: Have you seen the sitcom *The Big Bang Theory*? That's me: an astrophysicist. The way I like to explain my job is that it is to search for alien life. Not little green humanoids, but signs of life on planets orbiting other stars. Every star is a sun; each one has planets. There are endless suns with planets out there. We know of five thousand already, and there are, honestly, probably billions, or even trillions, of planets in our galaxy and universe. So the possibilities are huge and wondrous. In my memoir, I partially set out to share what I love about being a scientist. It's actually an exhilarating journey of exploration, and there are real ups and downs, just like everyday life. But there is also something very special about being on the cutting edge of discovery.

RHBC: This is a memoir of your whole life, and you write about your difficult, often lonely childhood. You were very close with your dad, but other than that you had a hard time making friends and building connections—until you met Mike. And it was many years later that you received your autism diagnosis. Can you talk about that?

SS: In one of my first memories as a child, I was in a carpool, coming home from kindergarten. Back in the day, kids all sat together without car seats in the back of a big station wagon. I just remember sitting there, and I saw the other girls in a semicircle, and I realized: wow. They are so connected with one another. And it wasn't a sad thought. It was just a realization that I was unable to connect with people in that way. Just recently, I got a formal diagnosis of being on the autism spectrum. And even though I was completely surprised, it made perfect sense. It explained my entire childhood. And why, even now, it takes so much effort to come across as "normal."

RHBC: It happened in an amazing way.

SS: Yes—the author of a profile about me in *The New York Times* is the parent of a child with autism, and apparently recognized it right away, but couldn't ask me outright. In the article, he wrote "wired in her own way." And that clicked into place for one of my mentors, a very good friend, whose wife is an autism expert. He explained that yes, this was of course why I could focus so well—I had more focus than anyone he's ever seen—and why I'm so good at my job. I just was in denial for the longest time.

RHBC: Would you say that your work provided a coping mechanism after Mike's death?

SS: Yes—both in helping me keep perspective and as an outlet for focus. There are trillions and trillions of stars in our universe. I love to imagine that there's an intelligent alien civilization on a planet orbiting a nearby star with the kind of telescope that we're trying to build. To them, our Earth—our whole planet—is just a pinprick of light, just another exoplanet. It was hard to confront my own personal tragedy against that vastness. So, through and through, it became my career goal to find another Earth and signs of life to show that we're not alone. But the Earths are so hard to find! They're so small and dim, right next to a big, bright, massive sun. I think a lot of us know what it's like to fall off the cliff to the abyss, to hit rock bottom. In that kind of depression, grief is its own category. But, actually, it allows new things to flourish. Like when a giant meteor catastrophically killed the dinosaurs, new life forms could grow. In my memoir, I aimed to capture the way that, sometimes, only because of huge disaster can an amazing, beautiful world emerge.

RHBC: You write about how your work helped you get through this difficult time, and you also write about how daily life—managing the household—was super challenging for you. Another powerful part of the book is the way Mike knew your head was always in the stars—not always here on Planet Earth—and his guide to life on Earth for you.

SS: Mike left me just three typed pages. Sometimes I laugh—he said I should hire a snow-removal service, and I thought, I can handle snow shoveling. But there I was, at 11:00 at night, shoveling, because the babysitter had pulled her shoulder and the kids were too little to help with snow. But it's not just that. With autism, I have to carefully think of everything before I say it, or it comes out wrong. And that's why I don't interact very well with the world around me. It was a really sweet thing that he did. With grief, you have no emotional reserves, so one way I navigated the world of widowhood was via accident, like when I met Melissa at the sledding hill.

RHBC: It's pretty spectacular to have met, in this small town, such a large group of young women who've lost their husbands. You got some dating advice—and outfits to borrow—from some of these friends. There's a lot of hope—even miracles—in this book about sadness, and, in the end, you do meet another man.

SS: It was just completely unexpected. I was a speaker at a conference, and at the reception, I walked into the room and saw a tall, dark, handsome man. I was so compelled to meet him. It was just like the movies, except it was an amateur astronomy conference. Meanwhile, Melissa, who often came with me to work events as my plus-one—she's really good at interacting with people—got to know about my work. I even took her on a trip to California, to a NASA center. And she at one point said, out of the blue, that it's harder to find true love than to find another Earth. I love to think about that, because it gives me confidence that I will find another Earth, because I found true love. The man at the conference,

Charles, and I have been married for about five years now, and he adopted my boys. My story is a journey of outer space, and the search for planets, and the exploration of our galaxy—but it's also about the journey of inner space, and the search that we all have to make our lives meaningful, especially after a major trauma.

RHBC: There are metaphors throughout the book about the ways that we're looking for connection and hope—looking for life. Some people say your professional mission is almost a folly, it seems like such an impossible task. Can you talk about what it's like to do this kind of science and work when it seems often against all odds?

SS: I know there are other worlds out there, and I am convinced there is life beyond Earth. It is hard for me to say I won't find another Earth or signs of life on another world, because I get up every day and I do my job. But the end of my book gets very philosophical—in that perhaps trying to find the so-called smallest lights in the universe isn't really exactly about who we'll encounter there. Maybe it isn't really about aliens and what they look like. Maybe it's more about us, and what the search says about us as humans. We're curious; we're hopeful. We're capable of wonder and wonderful things.

ABOUT THE AUTHOR

SARA SEAGER is an astrophysicist and a professor of physics and planetary science at MIT. Her research, which earned her a MacArthur Foundation "genius" grant, has introduced many foundational ideas to the field of exoplanets, and she led NASA's Probe-class study team for the Starshade project. She is now at the forefront of the search for the first Earth-like exoplanets and signs of life on them. She lives with her family in Concord, Massachusetts.

ABOUT THE TYPE

This book was set in Sabon, a typeface designed by the well-known German typographer Jan Tschichold (1902–74). Sabon's design is based upon the original letterforms of sixteenth-century French type designer Claude Garamond and was created specifically to be used for three sources: foundry type for hand composition, Linotype, and Monotype. Tschichold named his typeface for the famous Frankfurt typefounder Jacques Sabon (c. 1520–80).